ROAD TRANSPORT RESEARCH

ROAD MONITORING FOR MAINTENANCE MANAGEMENT

VOLUME 2

damage catalogue for developing countries

REPORT PREPARED BY AN OECD SCIENTIFIC EXPERT GROUP IN CO-OPERATION WITH THE WORLD BANK

The findings, interpretations, and recommendations in this study are the results of joint research by the OECD's Road Transport Research Programme and the World Bank. The report has been approved by the 23 governments participating in the Steering Committee of the OECD Road Transport Research Programme, i.e. all OECD Member governments except New Zealand. The report, however, does not necessarily reflect the views of the World Bank, its Board of Executive Directors, nor the countries they represent, other than those represented on the OECD RTR Steering Committee.

WORLD BANK
ORGANISATION FOR ECONOMIC CO-OPERATION AND DEVELOPMENT

Library of Congress Cataloging-in-Publication Data :

ROAD MONITORING MANUAL FOR MAINTENANCE MANAGEMENT.

Contents: Vol. 1. Manual for developing countries - Vol. 2. Damage catalogue for developing countries. 1. Roads - Developing countries - Maintenance and repair - Management. 2. Pavements - Developing countries - Maintenance and repair - Management.

I. ORGANISATION FOR ECONOMIC CO-OPERATION AND DEVELOPMENT. II. INTERNATIONAL BANK FOR RECONSTRUCTION AND DEVELOPMENT.

TE220.R594 1990 625.7'6'091724 89-70979
ISBN 92-64-13309-7 (set)
ISBN 0-8213-1391-6 (World Bank : set)

•

Pursuant to article 1 of the Convention signed in Paris on 14th December 1960, and which came into force on 30th September 1961, the Organisation for Economic Co-operation and Development (OECD) shall promote policies designed:

- to achieve the highest sustainable economic growth and employment and a rising standard of living in Member countries, while maintaining financial stability, and thus to contribute to the development of the world economy;
- to contribute to sound economic expansion in Member as well as non-member countries in the process of economic development; and
- to contribute to the expansion of world trade on a multilateral, non-discriminatory basis in accordance with international obligations.

The original Member countries of the OECD are Austria, Belgium, Canada, Denmark, France, the Federal Republic of Germany, Greece, Iceland, Ireland, Italy, Luxembourg, the Netherlands, Norway, Portugal, Spain, Sweden, Switzerland, Turkey, the United Kingdom and the United States. The following countries became Members subsequently through accession at the dates indicated hereafter: Japan (28th April 1964), Finland (28th January 1969), Australia (7th June 1971) and New Zealand (29th May 1973).

The Socialist Federal Republic of Yugoslavia takes part in some of the work of the OECD (agreement of 28th October 1961).

Publié en français sous le titre:

SUIVI DES ROUTES POUR LA GESTION DE L'ENTRETIEN
VOLUME 2
CATALOGUE DE DÉGRADATIONS
POUR LES PAYS EN DÉVELOPPEMENT

FOREWORD

The OECD Road Transport Research Programme has two main fields of activity:

-- The international co-operation in road and road transport research to provide scientific support for decisions by governments and international governmental organisations, and to assess future strategies concerning roads and road transport problems and the priority policy concerns of OECD Member countries;

-- The information and documentation programme (IRRD -- International Road Research Documentation), a co-operative scheme that provides a mechanism for the systematic world-wide exchange of information on scientific literature and current research programmes.

In order to assist development aid agencies and highway administrations in third world countries, one of the priorities of the Programme's scientific and technical activities is on the maintenance management of road infrastructure and the evaluation of traffic safety measures and strategies in developing countries.

This report contains a Road Damage Catalogue which is an essential tool for quantifying maintenance needs in developing countries. It is a complement to a larger study on road condition surveys and inspections published separately. The "Road Monitoring Manual" and the "Damage Catalogue" should help in improving maintenance practice and cost effectiveness of road maintenance operations in developing countries.

DEDICATION

This work is dedicated to Per Fossberg of the World Bank and David Brooks of the Overseas Unit of the U.K. Transport and Road Research Laboratory, both of whom died in 1989, in recognition of their outstanding contribution to international co-operation and to advancing road technology and maintenance in the developing world.

ABSTRACT

IRRD No. 826026

The OECD Scientific Expert Group on "Pavement Monitoring Systems for Developing Countries" was created to prepare, in co-operation with the World Bank, a manual for road pavement inspection applicable to Third World countries' conditions. The booklet contains a catalogue of the various types of road deficiencies most commonly found in developing countries. First it covers damage to be rated in road condition surveys relative to: (1) road surfacings, (2) drainage, shoulders and road space, (3) road signs and furnitures, as well as (4) structures. Secondly, the catalogue covers damage to be rated in detailed visual inspections of both (1) unpaved and (2) paved roads. For each type of damage the catalogue contains a sheet presenting (1) a description and definition, illustrated by a photo, suitable inspection methods and rating criteria and (2) information on probable damage causes and factors and possible remedies. The catalogue is a complement to the full report on "Road Monitoring for Maintenance Management: A Manual for Developing Countries".

Subject classification: 60

Fields: Maintenance

Keywords: Surveillance, manual (book), highway, developing countries, data acquisition, data processing, surfacing, evenness, damage, inventory, apparatus (measuring), evaluation (assessment), deterioration, road network

ACKNOWLEDGEMENTS

The Manual and Damage Catalogue were prepared by an OECD Road Transport Research Group with major input from the following experts: Messrs. A. Bresser and H. van den Beld (Netherlands), D. Brooks (United Kingdom), L. Francken (Belgium), M. Gorski (Belgium), H. Neubauer (Germany), W. Paterson (World Bank), C. Peyronne (France) and G. Refsdal (Chairman, Norway) and with the support of the OECD Secretariat and the World Bank. For the material received we should also like to thank the following institutions: Cadic Gombert, Consulting Engineers -- Belgium; Centre de Recherches Routières -- Belgium; DHV Consulting Engineers -- Netherlands; Forschungsgesellschaft für Strassen- und Verkehrswesen -- Germany; Laboratoire Central des Ponts et Chaussées -- France; Ministère des Transports et des Travaux Publics -- Mali; M & R International, Consulting Engineers -- Belgium; Overseas Unit, TRRL -- United Kingdom; Société d'ingénierie BCEOM -- France.

TABLE OF CONTENTS

		Page
I.	INTRODUCTION	7
II.	DAMAGE SHEETS	7
III.	RATING OF DAMAGE	9
	III.1. Road condition survey	9
	III.2. Detailed visual inspection	10
SECTION 1		
	ROAD CONDITION SURVEY: ROAD PARAMETERS	11
	Carriageway	11
	Shoulders	13
	Side drains	17
	Road space	21
	Road signs and furniture, road markings	25
SECTION 2		
	ROAD CONDITION SURVEY: STRUCTURES	29
	Culverts	31
	Bridges	37
SECTION 3		
	DETAILED VISUAL INSPECTION OF UNPAVED ROADS	45
	Rutting	47
	Corrugations	49
	Camber/crossfall	51
	Gravel thickness	53
	Erosion gullies	55
	Potholes	57
	Clay	59
SECTION 4		
	DETAILED VISUAL INSPECTION OF PAVED ROADS	61
	Rutting without cracks	63
	Rutting with cracks	65

Page

Corrugations 67
Depressions 69
Transverse cracks 71
Longitudinal cracks 73
Alligator cracks 75
Holes 77
Edge distress 79
Stripping/fretting/ravelling 81
Stripping/fretting/ravelling of surface 83
Bleeding 85

LIST OF GROUP MEMBERS 89

I. INTRODUCTION

To obtain evaluations as objective as possible and which are suitable for qualified personnel, it is mandatory to base visual inspections/surveys on a reference document such as a damage catalogue. The catalogue presented in this booklet comprises a number of sheets corresponding to those parameters which are evaluated and damage which most frequently occur in paved and unpaved roads. Figure 1 shows the position of the damage catalogue in the general scheme.

The proposed survey/inspection procedure includes two evaluation levels requiring a precise identification and quantification of principal defects. Although high output measuring equipment, such as the "Bump Integrator" or the "Longitudinal Profile Analyser" (APL) (see Annex D of the Manual), may be used for a quick inspection of the road system, this manual is based on visual surveys/ inspections, completed by simple measurements.

The damage is divided into that inspected during the Road Condition Survey (RCS) and that inspected during the Detailed Visual Inspection (DVI).

II. DAMAGE SHEETS

For each type of damage, the catalogue contains a sheet consisting of two parts:

A. The first part contains information relative to surveys/inspections:

- -- A photo or illustration if possible of the damage in various stages;
- -- A detailed description of the damage;
- -- A description of inspection and/or proposed measurement methods;
- -- Quantitative criteria to record the extent and/or severity of the damage.

B. The second part contains the following information:

- -- Probable cause of the damage, preliminary symptoms and consequences, if no corrective maintenance actions are carried out;

Figure 1. **POSITION OF THE DAMAGE CATALOGUE IN THE GENERAL SCHEME**

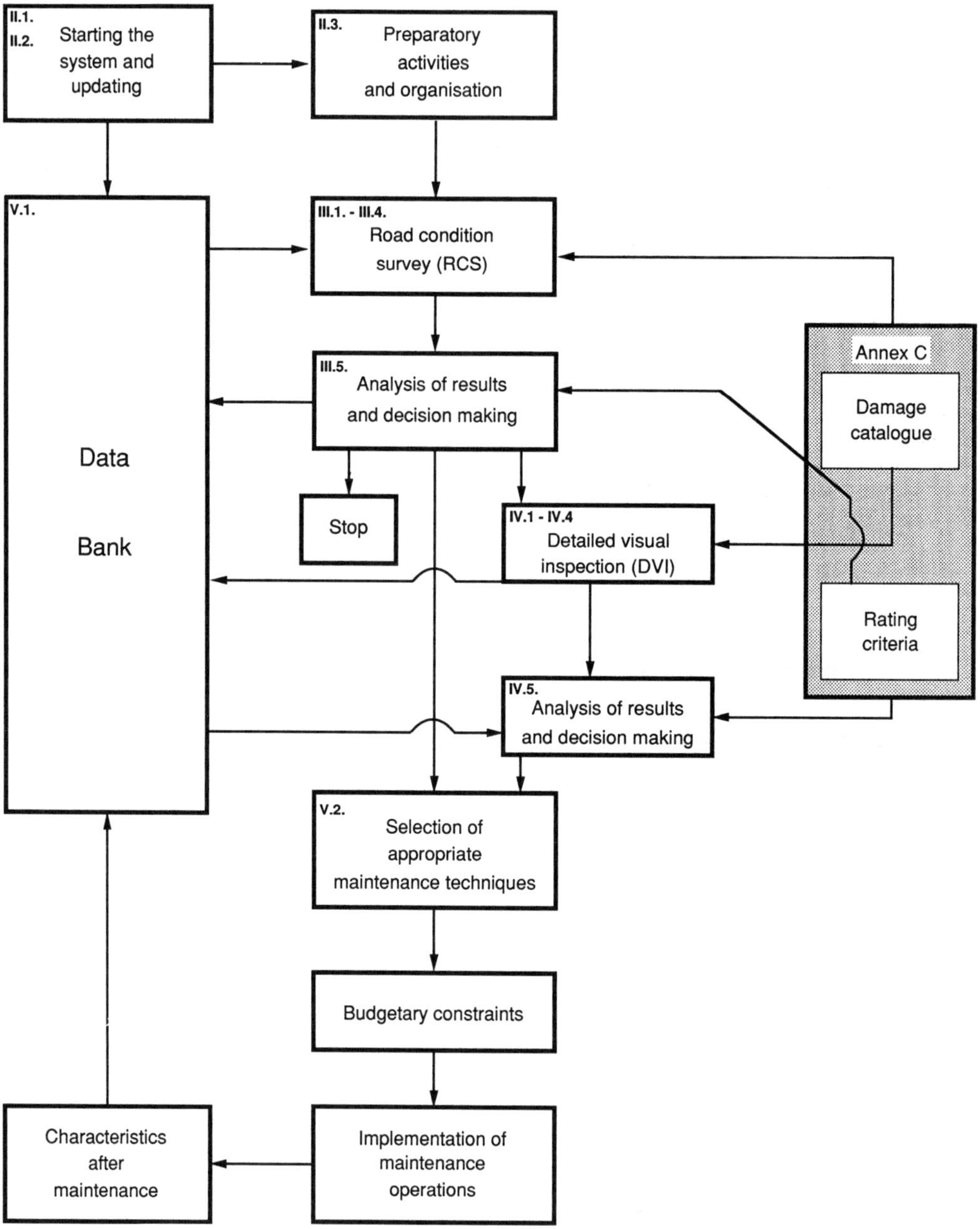

-- Factors affecting the development of the damage such as climate, traffic and materials;

-- Proposed operations or remedies (divided into specific tasks), and taking into account the various causes leading to the present situation.

The last section is indicative only and the information given is summarised and incomplete. As such it should not be used by the maintenance department to define the maintenance activity to be undertaken. Other publications deal more specifically with maintenance and rehabilitation techniques and it is recommended to use these.

III. RATING OF DAMAGE

III.1. Road Condition Survey (RCS)

The classification of the carriageway or road surface condition is done on a five-point scale, based upon the criteria given in Annex B of the full Manual.

In addition to the damage affecting the road surfacing, the inspectors also evaluate the defects and damage of the drainage system and the area adjacent to the road (fill, etc.). This is done with a three-point rating system for roadside elements, road signs and furniture and structures in accordance with the description given on the relevant damage sheet.

The principles of this three-point system are that the values recorded for each element reflect the efficiency of routine and recurrent maintenance activities. The need for immediate intervention is based on considerations relating to road user safety and the stability of the road, embankment and structures. It is recorded in the "Remarks" section.

RATING CRITERIA

Value	Routine maintenance	Action
1	Satisfactory	Not necessary
2	To be checked	To be expected shortly
3	Not satisfactory	Necessary

III.2. Detailed Visual Inspection (DVI)

The rating of damage affecting the carriageway is based on two characteristics: extent and severity.

The extent may be defined as that part of the road affected by the damage.

For linear damage (rutting, edge distress, longitudinal cracks) the inspector estimates the length of the damage in proportion to the length of the sub-section.

For two-dimensional damage (corrugations, alligator cracking, peeling, bleeding, etc.) the inspector estimates the damaged area in proportion to the total area of the sub-section.

For both cases the proposed method distinguishes only three levels. These levels differ for the various types of defects and are given on the relevant sheet.

In this method, the severity of the damage is expressed as:

1. light;
2. moderate;
3. severe.

Purely superficial damage such as deposits of clay (CY) on unpaved roads or bleeding (BL) on paved roads cannot be quantified in terms of severity.

The significance of the damage is derived from a combination of the extent and severity. This is rated on a five-point scale established in matrices presented on the damage sheets. Apart from the defects CY and BL, all defects listed on Forms IV and V are assigned a value in accordance with the five-point rating scale. The defects CY and BL are assigned a value on the basis of their extent.

Note:

-- It is of the utmost importance that the inspection teams have an intimate knowledge of the information contained in the damage catalogue.

-- It is strongly recommended that, prior to actual inspection work, teams receive proper training related to the inspection programme (see Chapter VI of the full Manual).

SECTION 1

ROAD CONDITION SURVEY: ROAD PARAMETERS

Group	Type	No. of sheet	Rating levels	Section
Carriageway	Surface quality	-	5	Annex B
	Prevailing damage (if necessary)	-	-	3-4
Shoulders	Deformation	RCS 1	3	1
	Scour	RCS 2	3	
Side drains	Siltation	RCS 3	3	1
	Scour	RCS 4	3	
Road space	Debris/vegetation encroachment	RCS 5	3	1
	Obstacles	RCS 6	3	
Road signs & furniture; Road markings	Damaged and/or dirty, missing	RCS 7	3	1
Damage & defects to be repaired immediately	Various, to be mentioned under "Remarks"	-	-	3-4

The rating of the carriageway or road surface condition is based upon the criteria given in Annex B of the full report.

Figure 2. **ROAD CONDITION SURVEY (RCS) FORM ("FORM II")**

ROAD CONDITION SURVEY PAVED AND UNPAVED ROADS		ROAD DENOMINATION / CLASS:	Main Section No.
		FROM:	Km:
		TO:	Km:
OFFICE:	DISTRICT:	Pavement type: / Pavement width: m	Section length: Km
Name of inspector:		Summary (Avg. Condition)	Carriageway:
			Road side elements:
Date:	Weather o Clear o Rainy / Carr.way: o Dry o Drying o Wet		Remarks:

SUB-SECTION	◯	◯	◯	◯	◯
Chainage of sub-section					

Roadside	L	R	L	R	L	R	L	R	L	R
Carriageway	1 2 4 3 5	1 2 4 3 5	1 2 4 3 5	1 2 4 3 5	1 2 4 3 5	1 2 4 3 5	1 2 4 3 5	1 2 4 3 5	1 2 4 3 5	1 2 4 3 5
Prevailing damage										

Damage type			L	R	L	R	L	R	L	R	L	R
ROAD SIDE ELEMENTS	Shoulder	Deformation RCS1	1 2 3	1 2 3	1 2 3	1 2 3	1 2 3	1 2 3	1 2 3	1 2 3	1 2 3	1 2 3
		Scour RCS2	1 2 3	1 2 3	1 2 3	1 2 3	1 2 3	1 2 3	1 2 3	1 2 3	1 2 3	1 2 3
	Side drains	Siltation RCS3	1 2 3	1 2 3	1 2 3	1 2 3	1 2 3	1 2 3	1 2 3	1 2 3	1 2 3	1 2 3
		Scour RCS4	1 2 3	1 2 3	1 2 3	1 2 3	1 2 3	1 2 3	1 2 3	1 2 3	1 2 3	1 2 3
	Debris/Vegetation Encroachment RCS5		1 2 3	1 2 3	1 2 3	1 2 3	1 2 3	1 2 3	1 2 3	1 2 3	1 2 3	1 2 3
	Obstacles/ Obstructions RCS6		1 2 3	1 2 3	1 2 3	1 2 3	1 2 3	1 2 3	1 2 3	1 2 3	1 2 3	1 2 3
			1 2 3	1 2 3	1 2 3	1 2 3	1 2 3	1 2 3	1 2 3	1 2 3	1 2 3	1 2 3
			1 2 3	1 2 3	1 2 3	1 2 3	1 2 3	1 2 3	1 2 3	1 2 3	1 2 3	1 2 3
	Average Condition (Others)		+ =		+ =		+ =		+ =		+ =	
			: =		: =		: =		: =		: =	
ROADSIGNS/FURNITURE	Chainage											
	Dirty RCS7		1 2 3	1 2 3	1 2 3	1 2 3	1 2 3	1 2 3	1 2 3	1 2 3	1 2 3	1 2 3
	Damaged RCS7		1 2 3	1 2 3	1 2 3	1 2 3	1 2 3	1 2 3	1 2 3	1 2 3	1 2 3	1 2 3
	Missing RCS7		1 2 3	1 2 3	1 2 3	1 2 3	1 2 3	1 2 3	1 2 3	1 2 3	1 2 3	1 2 3
			1 2 3	1 2 3	1 2 3	1 2 3	1 2 3	1 2 3	1 2 3	1 2 3	1 2 3	1 2 3

REMARKS: ______________________________

OECD DC1	DAMAGE SHEET	RCS 1
GROUP: SHOULDER	TYPE: DEFORMATION	

Rising shoulder due to spreading vegetation with localised road surface bumps.

Lowered shoulder due to erosion. Starting deformation of road surface edges.

DESCRIPTION:

The visual aspects may differ according to the type of deformation. Two cases are distinguished:
Case a : the shoulder level is higher than the road surface level.
Case b : the shoulder level is lower than the road surface level.

INSPECTION METHOD:

Visual recording and estimation of length of the deformed shoulder expressed as a percentage of the sub-section length.

RATING CRITERIA:

Value	Extent	Routine maintenance	Action
1	< 5 %	Satisfactory	Not required
2	5-50 %	To be checked	To be expected
3	> 50 %	Not satisfactory	Required immediately

OECD DC1	DAMAGE SHEET	RCS 1

GROUP:	SHOULDER	TYPE:	DEFORMATION

PROBABLE CAUSES AND FACTORS AFFECTING THE DEVELOPMENT OF THE DAMAGE:

Case a:

-- Accumulation of material from the road surface due to traffic and displacement by water
-- Growth of grasses and bushes on unpaved shoulders
-- Slope failures affecting the shoulder

Case b:

-- Wear and deformation of shoulders caused by traffic and parking on shoulders in particular at soft spots and less compacted areas
-- Strengthening of carriageway
-- Slope failure of side drain due to erosion, water soaked embankment or embankment erosion
-- Unforeseen settlement of embankment and subsequent damage of side drain and shoulder

REMARKS:

Following are the consequences of deformed shoulders:

Case a:

Ponding and infiltration of water on the side of the carriageway due to improper drainage of surface water; deformation of carriageway.

Case b:

Ponding and infiltration of water on shoulders beside the carriageway. Deformation of carriageway; edge damage risks due to vehicle wheels. Dangerous for traffic.

In all cases rutting (PR 1b) and depressions (UR 3) will be generated.

POSSIBLE MAINTENANCE ACTIVITY:

-- Reshape the shoulders to a lower level.
-- Vegetation growth control.
-- Protect slopes (in cut) against erosion.
-- Refill shoulders.
-- Make or repair outlets for side drains.
-- Stop surface drainage over fill slopes and take protective measures against erosion.

OECD DC1	DAMAGE SHEET	RCS 2

GROUP: SHOULDER TYPE: SCOUR

Generalised erosion of shoulder

Erosion of shoulder and embankment

DESCRIPTION:

Two different types can be distinguished:

a) Gullies perpendicular to the road axis cutting through the full width of the shoulder
b) Longitudinal gullies, parallel to the road axis.

INSPECTION METHOD:

Visual recording and estimation of the eroded length expressed as a percentage of the sub-section length.

RATING CRITERIA:

Value	Extent	Routine maintenance	Action
1	< 5 %	Satisfactory	Not required
2	5-50 %	To be checked	To be expected
3	> 50 %	Not satisfactory	Required immediately

OECD DC1	DAMAGE SHEET	RCS 2

GROUP:	SHOULDER	TYPE:	SCOUR

PROBABLE CAUSES AND FACTORS AFFECTING THE DEVELOPMENT OF THE DAMAGE:

Erosion due to run-off of surface water.
The defect might be made worse by the flooding of the shoulder due to flooding of the side drain.
Main factors for the development of the defect are the erosive action of rain; the corresponding washing out of erodable soils; depends also on the topographical conditions.
The defect appears in particular on steep slopes, at curves in badly compacted materials and in easily erodable materials.
Gullies will be aligned along the steepest slope of the shoulder embankment.

REMARKS:

Progressive scour deepens the gullies and the shoulder becomes unusable. This endangers the safety of the road users.

POSSIBLE MAINTENANCE ACTIVITY:

1-2. Reshape and relevel the shoulder.
3. In case of extensive erosion reshape the shoulder by adding well compacted erosion resistant material.

OECD DC1	DAMAGE SHEET	RCS 3
GROUP: SIDE DRAINS	TYPE: SILTATION	

Side drains filled up by fine particles due to erosion and water run-off.

Slope erosion resulting in ditch siltation.

DESCRIPTION:

Deposit of material in the side drains of the road causing a partial or completely blocked drainage system.

INSPECTION METHOD:

Visual recording and estimation of the total length of the affected drains as a percentage of the sub-section length.

RATING CRITERIA:

Value	Extent	Routine maintenance	Action
1	< 5 %	Satisfactory	Not required
2	5-50 %	To be checked	To be expected
3	> 50 %	Not satisfactory	Required immediately

OECD DC1	DAMAGE SHEET	RCS 3

GROUP:	SIDE DRAINS	TYPE:	SILTATION

PROBABLE CAUSES AND FACTORS AFFECTING THE DEVELOPMENT OF THE DAMAGE:

a) Presence of obstacles, resulting from:
   - -- Side drain slope failure due to traffic action or cattle movement; or to erosion; or to cut slope failure;
   - -- Encroaching vegetation and accumulation of debris in drains.

b) Bad drainage design:
   - -- Too shallow longitudinal slope in the side drain;
   - -- Insufficient outlets to disperse the water.

REMARKS:

The consequences of the siltation are:

-- Reduction of ditch cross-section and velocity of water;
-- Increasing siltation and amounts of debris resulting in a complete obstruction of the ditch itself and the outlets;
-- Ponding of water in the ditch, water penetration in base and sub-base course resulting in erosion;
-- Reduction of bearing capacity of road shoulder.

POSSIBLE MAINTENANCE ACTIVITY:

-- Cleaning up and reshaping of drains, control vegetation growth;
-- Cleaning up and repair of catchment basins;
-- Cleaning up and increasing number of outlets;
-- Eventual construction of interception ditches to improve erosion from slopes.

OECD DC1	DAMAGE SHEET	RCS 4
GROUP: SIDE DRAINS	TYPE: SCOUR	

Regressive erosion of a side drain.

Erosion of a turnout.

DESCRIPTION:

Erosion of bottom and sides of ditches:
- -- Direct erosion is occurring when soil particles are loosened and transported by water;
- -- Regressive erosion is occurring when a relatively flat terrain is followed (downstream) by a terrain with a steeper slope or more erodable materials.

INSPECTION METHOD:

Visual recording and estimation of the total length of the eroded drains as a percentage of the sub-section length.

RATING CRITERIA:

Value	Extent	Routine maintenance	Action
1	< 5 %	Satisfactory	Not required
2	5-50 %	To be checked	To be expected
3	> 50 %	Not satisfactory	Required immediately

OECD DC1	DAMAGE SHEET	RCS 4

GROUP:	SIDE DRAINS	TYPE:	SCOUR

PROBABLE CAUSES AND FACTORS AFFECTING THE DEVELOPMENT OF THE DAMAGE:

-- Use of non cohesive materials which are not stabilized;
-- Too steep longitudinal slope of drain;
-- Drain too long (superior to critical length) insufficient number of outlets;
-- Irregular longitudinal profile of drain; sudden change of slope; level of outlet too low.

REMARKS:

Scour in the drain will result in the collapse of the drain in the first instance; rapidly the shoulder will follow and ultimately the carriageway itself will be endangered. Regressive erosion provokes a number of successive embankment collapses which endanger the road user and furthermore leads to major repair costs.

POSSIBLE MAINTENANCE ACTIVITY:

-- Reshape and realign ditch;
-- Stabilise or protect bottom and side slopes of ditch in sensitive areas;
-- Reduce ditch length, construct more outlets;
-- Construct sedimentation dams.

OECD DC1	DAMAGE SHEET	RCS 5

GROUP: ROAD SPACE TYPE: DEBRIS/VEGETATION ENCROACHMENT

Dangerous crossroad due to
accumulating mud.

DESCRIPTION:

Obstruction of the carriageway which will hinder or endanger the traffic.

INSPECTION METHOD:

Visual recording and estimation of the severity of the obstruction.

RATING CRITERIA:

Value	Severity	Routine maintenance	Action
1	No major obstacles	Satisfactory	Not required
2	Necessary to slow down and bypass an obstacle. Major visibility reduction.	To be checked	To be expected
3	Total obstruction	Not satisfactory	Required immediately

OECD DC1	DAMAGE SHEET	RCS 5

GROUP:	ROAD SPACE	TYPE:	DEBRIS/VEGETATION ENCROACHMENT

PROBABLE CAUSES AND FACTORS AFFECTING THE DEVELOPMENT OF THE DAMAGE:

Obstacles might result from:

-- Sand encroachment following a storm or dune displacement;
-- Vegetation encroachment due to insufficient maintenance of grasses and bushes alongside the road, or cut or fallen trees;
-- Slope failures due to bad design or due to water penetration in slopes.

REMARKS:

Possible consequences:

-- Reduction of visibility distances and increase of accident risks;
-- Accumulation of material on the side of the carriageway blocking surface drainage;
-- Increased surface deterioration due to trafficking of the resultant narrow carriageway;
-- Enforced use of shoulders by vehicles.

POSSIBLE MAINTENANCE ACTIVITY:

-- Reduce vegetation growth, cleaning and grubbing of shoulders, cutting trees along the road alignment;
-- Clearing the roadway of vegetation and sand (dunes);
-- Repairing collapsed slopes.

GROUP: ROAD SPACE TYPE: OBSTACLES/OBSTRUCTIONS

Obstructed carriageway due to unstable side slope.

DESCRIPTION:

Obstruction of the carriageway which will hinder or endanger traffic flow.

INSPECTION METHOD:

Visual recording and estimation of severity and effect on traffic flow.

RATING CRITERIA:

Value	Severity	Routine maintenance	Action
1	No major obstacles	Satisfactory	Not required
2	Necessary to slow down and bypass an obstacle. Major visibility reduction.	To be checked	To be expected
3	Total obstruction	Not satisfactory	Required immediately

OECD DC1	DAMAGE SHEET	RCS 6

GROUP:	ROAD SPACE	TYPE:	OBSTACLES/OBSTRUCTIONS

PROBABLE CAUSES AND FACTORS AFFECTING THE DEVELOPMENT OF THE DAMAGE:

Obstacles may result from:

-- Repair of vehicles (leaving behind tools or other items) on the road;
-- Broken down vehicles;
-- Rock fall.

REMARKS:

Possible consequences:

-- Increase of accident risks to traffic;
-- Enforced use of shoulders.

POSSIBLE MAINTENANCE ACTIVITY:

-- Removal of obstacles from the carriageway;
-- Provision of appropriate traffic/warning signs when obstacle cannot immediately be removed.

OECD DC1	DAMAGE SHEET	RCS 7

GROUP:	ROAD SIGNS & FURNITURE ROAD MARKING	TYPE:	DAMAGED/MISSING

Accidental destruction of a bridge railing.

DESCRIPTION:

Damage to road signs, guard rails, paint markings, etc.; road signs, guardrails or markings are missing at critical places, e.g. road crossings or hair-pin curves, ravines, etc.

INSPECTION METHOD:

Record for each section all signs, furniture and markings and estimate the number of missing elements as a percentage of the total number.
Describe the location and severity of damaged elements and implement immediate repairs or replacements.

RATING CRITERIA:

Value	Extent
1	< 10 %
2	10-25 %
3	> 25 %

OECD DC1	DAMAGE SHEET	RCS 7

GROUP:	ROAD SIGNS & FURNITURE ROAD MARKING	TYPE:	DAMAGED/MISSING

PROBABLE CAUSES AND FACTORS AFFECTING THE DEVELOPMENT OF THE DAMAGE:

-- Wear and tear by traffic;
-- Damage from traffic accidents;
-- Weather (corrosion);
-- Vandalism.

REMARKS:

Damaged signs and road furniture endanger traffic safety and should therefore be repaired quickly. (This may also apply to dirty road signs.) For road safety reasons, missing signs, road markings and furniture have to be replaced immediately.

POSSIBLE MAINTENANCE ACTIVITY:

-- Repair or replacement of damaged (missing) elements;
-- Repainting of road markings.

Figure 3. **SUMMARY OF ROAD CONDITION SURVEY RESULTS: AVERAGE CONDITION VALUE ("FORM IV")**

ROAD CONDITION SURVEY (RCS)

PAVED AND UNPAVED ROADS

OFFICE:	DISTRICT:	PAGE
Established by:		
Date:		

DESIGNATION OF ROAD SECTION								
Road		Section No.	Identification Points (IP)			Section length	Pavement width	AADT
Class	No.		I.P.	Designation	Km	Km	m	Veh/day
1	2	3	4	5	6	7	8	9
			From To		From To			
			From To		From To			
			From To		From To			
			From To		From To			
			From To		From To			
			From To		From To			
			From To		From To			
			From To		From To			
			From To		From To			
			From To		From To			

SURVEY DATA			
Avg. condition value			Notes / Remarks
Carriageway	Road side elements	Road signs & Furniture	
11	12	13	14

SECTION 2

ROAD CONDITION SURVEY: STRUCTURES

Group	Type	No. of sheet	Rating levels
Culverts	Silted/blocked	RCS 8	3
	Scour	RCS 9	3
	Structural damage	RCS 10	3
Bridges	Silted/blocked	RCS 11	3
	Scour	RCS 12	3
	Structural damage	RCS 13	3

Figure 4. **CONDITION SURVEY FORM FOR STRUCTURES ("FORM III")**

ROAD CONDITION SURVEY (RCS)

OFFICE:		DISTRICT:	PAGE
Name of inspector:			
Date:	Weather o Clear o Rainy		

STRUCTURES

Road denomination/Class:

Type of structure	Main section	Sub-section	Chainage

CONDITION SURVEY

Silted/ Blocked		Scour		Structural Damage		Notes/Remarks
L	R	L	R	L	R	
1 2 3	1 2 3	1 2 3	1 2 3	1 2 3	1 2 3	
1 2 3	1 2 3	1 2 3	1 2 3	1 2 3	1 2 3	
1 2 3	1 2 3	1 2 3	1 2 3	1 2 3	1 2 3	
1 2 3	1 2 3	1 2 3	1 2 3	1 2 3	1 2 3	
1 2 3	1 2 3	1 2 3	1 2 3	1 2 3	1 2 3	
1 2 3	1 2 3	1 2 3	1 2 3	1 2 3	1 2 3	
1 2 3	1 2 3	1 2 3	1 2 3	1 2 3	1 2 3	
1 2 3	1 2 3	1 2 3	1 2 3	1 2 3	1 2 3	
1 2 3	1 2 3	1 2 3	1 2 3	1 2 3	1 2 3	
1 2 3	1 2 3	1 2 3	1 2 3	1 2 3	1 2 3	
1 2 3	1 2 3	1 2 3	1 2 3	1 2 3	1 2 3	
1 2 3	1 2 3	1 2 3	1 2 3	1 2 3	1 2 3	
1 2 3	1 2 3	1 2 3	1 2 3	1 2 3	1 2 3	

OECD DC1	DAMAGE SHEET	RCS 8
GROUP: CULVERTS	TYPE: SILTED/BLOCKED	

Partial obstruction due to progressive siltation.

DESCRIPTION:

This type of damage concerns the partial or complete obstruction of culverts by silt, natural debris or other material.

INSPECTION METHOD:

Visual inspection of all sub-section culverts and evaluation of functional conditions.

RATING CRITERIA:

Value	Severity	Routine maintenance	Action
1	Free flow	Satisfactory	Not required
2	Partial obstruction	To be checked	To be expected
3	Total obstruction	Not satisfactory	Required immediately

OECD DC1	DAMAGE SHEET	RCS 8
GROUP: CULVERTS	TYPE: SILTED/BLOCKED	

PROBABLE CAUSES AND FACTORS AFFECTING THE DEVELOPMENT OF THE DAMAGE:

Design faults:

-- Invert level too low;
-- Longitudinal slope too flat;
-- Culvert pipe diameter incorrect.

During rainy seasons, soil/sand, small trees and vegetation in ground is carried away by the rain water and deposit in initial locations (entrance of culvert pipe, badly designed structures); this will lead to obstructions and subsequently flooding and inundation.

REMARKS:

Lack of inspection and maintenance may have the following consequences:

-- Submersion of embankment resulting in uncontrolled settlement of road shoulders;
-- Ponding of water upstream of the structure resulting in slope failure;
-- Inundation of complete culvert, shoulders and road.

POSSIBLE MAINTENANCE ACTIVITY:

-- Preventative maintenance in dry season;
-- Cleaning of culverts especially entrance and mid-section of culvert pipes;
-- Construction of protective structures, against silt and debris upstream of culvert (mesh, deflecting devices, etc.);
-- Correction of culvert design errors.

GROUP: CULVERTS TYPE: SCOUR

Scour at the end of the culvert.

DESCRIPTION:

This type of damage consists of the loss of materials from both ends of the culvert; eventually the whole structure might be undermined. Two types of erosion can take place downstream of the culvert depending on the initial gradient:

-- Localised erosion with gully development;
-- Regressive erosion with ravine development.

INSPECTION METHOD:

Visual inspection of all sub-section culverts and evaluation of severity of scour.

RATING CRITERIA:

Value	Severity	Routine maintenance	Action
1	Little or no erosion	Satisfactory	Not required
2	Substantial erosion without settlement of structure	To be checked	To be expected
3	Substantial erosion undermining the structure	Not satisfactory	Required immediately

OECD DC1	DAMAGE SHEET	RCS 9

GROUP:	CULVERTS	TYPE:	SCOUR

PROBABLE CAUSES AND FACTORS AFFECTING THE DEVELOPMENT OF THE DAMAGE:

The severity of the localised erosion and the damage caused by it depend upon the conditions at the downstream end of the culvert (flow, diameter, Froude parameter).

Regressive erosion with ravine development depends upon the stability of the free flow and the downstream outlet conditions of the culvert.

REMARKS:

The erosion will ultimately attack the embankment slopes and causes damage to the complete structure such as head and wingwalls and the culvert itself.

POSSIBLE MAINTENANCE ACTIVITY:

Design and implement protective erosion control and energy dissipating structures.

-- Backfill of scoured areas with boulders or gabions;
-- Installation of gabion protection, rock beds, etc.;
-- Construction of concrete protection structures.

OECD DC1	DAMAGE SHEET	RCS 10

GROUP:	CULVERTS	TYPE:	STRUCTURAL DAMAGE

Total destruction of a gully with collapse of the pavement

DESCRIPTION:

Two types of structural damage can be distinguished:

-- Slight settlement with subsequent cracking in head walls, wing walls and the culvert structure;

-- Pronounced uneven settlement, with the subsequent development of large cracks.

INSPECTION METHOD:

Visual inspection of all sub-section culverts and evaluation of severity of damage.

RATING CRITERIA:

Value	Severity	Routine maintenance	Action
1	Small or no damage	Satisfactory	Not required
2	Uneven settlement damaged inlet and outlets	To be checked	To be expected
3	Broken culvert (pipes), inlet and outlets	Not satisfactory	Required immediately

OECD DC1	DAMAGE SHEET	RCS 10

GROUP:	CULVERTS	TYPE:	STRUCTURAL DAMAGE

PROBABLE CAUSES AND FACTORS AFFECTING THE DEVELOPMENT OF THE DAMAGE:

-- Incorrect design;
-- Diminished bearing capacity of foundation layers, settlement of soils under foundation;
-- Erosive action around structure;
-- Traffic may cause damage when insufficient cover is present.

REMARKS:

Lack of inspection and maintenance may have the following consequences:

-- Minor damage : culvert is still functional;
-- Major damage : obstruction of culvert, collapse of structure.

POSSIBLE MAINTENANCE ACTIVITY:

-- Repair cracks;
-- Reconstruct inlet and outlet and other missing parts;
-- Reconstruct complete structure.

OECD DC1	DAMAGE SHEET	RCS 11
GROUP: BRIDGES	TYPE: SILTED/BLOCKED	

Accumulation of debris and vegetation below a bridge.

Shrubs which may lead to obstruction.

DESCRIPTION:

Depending on the nature of the structure, the following types of damage can be distinguished:

-- Accumulation of soil or debris on the bridge deck and in the deck joints;
-- Accumulation of floating debris against piers, abutments.

INSPECTION METHOD:

Visual inspection of all sub-section bridges and evaluation of severity of defect.

RATING CRITERIA:

Value	Severity	Routine maintenance	Action
1	Clean bridge deck, free flow under bridge	Satisfactory	Not required
2	Debris accumulation on and under bridge	To be checked	To be expected
3	Debris obstructing traffic flow and water flow	Not satisfactory	Required immediately

GROUP: BRIDGES TYPE: SILTED/BLOCKED

PROBABLE CAUSES AND FACTORS AFFECTING THE DEVELOPMENT OF THE DAMAGE:

During rainy seasons:

-- Temporary inundation, flooding of bridge deck, malfunctioning of discharge facilities;
-- Transport of floating debris in the river during periods of high flow; insufficient free flow under bridge.

Design deficiencies:

-- Insufficient bridge width/span;
-- Piers/abutments obstructing the flow of water.

REMARKS:

Lack of maintenance may have the following consequences:

Bridge deck	: slippery surface endangering traffic;
Joints	: accumulation of dirt in joints will obstruct the free expansion of the deck;
Abutments	: obstruction of riverbed, reducing outflow, increasing velocity of water and scour risks.

POSSIBLE MAINTENANCE ACTIVITY:

-- Clean discharge facilities, repair immediately if not functional;
-- Clean joints;
-- Remove all floating dirt from piers and abutments;
-- If the bridge opening is obstructed regularly, consider widening.

OECD DC1	DAMAGE SHEET	RCS 12

GROUP: BRIDGES TYPE: SCOUR

Sagging of abutment and paved side slopes resulting in structural damage.

Collapse of bridge embankment uncovering the wall of the abutment and undermining the pavement structure.

DESCRIPTION:

The following types of damage can be distinguished:

-- Erosion of slopes and bedding;
-- Undermining of the abutments and piers.

INSPECTION METHOD:

Visual inspection of all sub-section bridges and estimate of severity of damage. Inspect slopes, bedding, walls, rock beds, piers next to the bridge.

RATING CRITERIA:

Value	Severity	Routine maintenance	Action
1	Little or no erosion	Satisfactory	Not required
2	Erosion without major damage	To be checked	To be expected
3	Severe erosion endangering road and bridge	Not satisfactory	Required immediately

OECD DC1	DAMAGE SHEET	RCS 12
GROUP: BRIDGES	TYPE: SCOUR	

PROBABLE CAUSES AND FACTORS AFFECTING THE DEVELOPMENT OF THE DAMAGE:

-- No discharge facilities for run off of surface water next to the bridge, infiltration of water between bridge and approach road, instability of slopes.
-- Increase of flow velocity due to narrowing or obstruction of river bed.
-- Piers/abutments not in line with flow of water.

REMARKS:

Lack of inspection and maintenance may have the following consequences:

-- Slope failures, unstable abutments, ramps between approach road and bridge deck endangering traffic;
-- Uneven settlement of foundations, collapse of bridge.

POSSIBLE MAINTENANCE ACTIVITY:

-- Repair scoured slopes;
-- Protect eroded areas such as piers/abutments and slopes with boulders, gabions or retaining walls.

OECD DC1	DAMAGE SHEET	RCS 13
GROUP: BRIDGES	TYPE: STRUCTURAL DAMAGE	

Partial destruction of wooden bridge deck.

Exposure of steel reinforcement of bridge deck.

DESCRIPTION:

The following types of damage can be distinguished:

-- Degradation of deck;
-- Degradation of masonry joints;
-- Cracking of abutments;
-- Exposure of reinforcement steel.

INSPECTION METHOD:

Visual inspection of all sub-section bridges and estimate of severity of damage.

RATING CRITERIA:

Value	Severity	Routine maintenance	Action
1	Small or no damage	Satisfactory	Not required
2	Visible damage without risk to the bridge or traffic	To be checked	To be expected
3	Visible damage endangering the bridge and traffic	Not satisfactory	Required immediately

GROUP: BRIDGES TYPE: STRUCTURAL DAMAGE

PROBABLE CAUSES AND FACTORS AFFECTING THE DEVELOPMENT OF THE DAMAGE:

Bridge deck	:	loosened bolts, anchor bolts, nails; wear and tear;
Abutments	:	aggressive water action on joints, weathering action of climate (corrosive action), settlement of foundation soils, increasing soil pressure;
Supports and joints	:	blockage due to dirt and corrosive action of climate leading to damage to abutment and bridge deck.

REMARKS:

Lack of inspection and maintenance may have the following consequences:

-- Loss of bridge deck elements, damage to tyres and vehicles;
-- Localised collapse of masonry, settlement of abutments, complete bridge collapse.

POSSIBLE MAINTENANCE ACTIVITY:

-- Maintain and clear bridge deck and abutments;
-- Repair or replace deck;
-- Repair damaged masonry, protect degraded/corroded/cracked areas;
-- Rebuild bridge in the case of severe settlement or risk of collapse.

Figure 5. **SUMMARY OF SURVEY RESULTS FOR STRUCTURES : AVERAGE CONDITION VALUE PER STRUCTURE ("FORM V")**

ROAD CONDITION SURVEY (RCS)

OFFICE: | DISTRICT:

Established by:

Date:

PAGE

STRUCTURES				CONDITION SURVEY DATA: AVERAGE CONDITION VALUE			
Road denomination/Class:							
Type of structure	Main section	Sub-section	Chainage	Silted/ Blocked	Scour	Structural damage	Notes / Remarks

SECTION 3

DETAILED VISUAL INSPECTION OF UNPAVED ROADS

Type	No. of sheet	Code	Function affected	Rating levels
Rutting	DVI/UR 1	RU	Roughness/ Drainage	5
Corrugations	DVI/UR 2	CO	Roughness	5
Camber/crossfall	DVI/UR 3	CR	Drainage	5
Gravel thickness	DVI/UR 4	GR	Structure	5
Erosion gullies	DVI/UR 5	EG	Structure	5
Potholes	DVI/UR 6	PO	Structure	5
Clay	DVI/UR 7	CY	Safety	3

Figure 6. **DETAILED VISUAL INSPECTION FORM FOR UNPAVED ROADS ("FORM VI")**

DETAILED VISUAL INSPECTION (DVI)			**UNPAVED ROADS**		
OFFICE:	DISTRICT:		ROAD DENOMINATION / CLASS:		Main Section No.
Name of inspector:			FROM I.P.:		Km:
Date:	Weather o Clear o Rainy	Carr.way: o Dry o Drying o Wet	TO I.P.:		Km:
			Pavement type:	Pavement width: m	Section length: Km

SUB-SECTION		○						○						○						○						○					
Roadside		L			R			L			R			L			R			L			R			L			R		
Damage type - Cat	E \ S	L	M	S	L	M	S	L	M	S	L	M	S	L	M	S	L	M	S	L	M	S	L	M	S	L	M	S	L	M	S
Rutting - UR1	<10%	1	3	5	1	3	5	1	3	5	1	3	5	1	3	5	1	3	5	1	3	5	1	3	5	1	3	5	1	3	5
	10-50%	2	4	5	2	4	5	2	4	5	2	4	5	2	4	5	2	4	5	2	4	5	2	4	5	2	4	5	2	4	5
	>50%	3	5	5	3	5	5	3	5	5	3	5	5	3	5	5	3	5	5	3	5	5	3	5	5	3	5	5	3	5	5
Corrugations - UR2	<10%	1	3	4	1	3	4	1	3	4	1	3	4	1	3	4	1	3	4	1	3	4	1	3	4	1	3	4	1	3	4
	10-50%	2	3	5	2	3	5	2	3	5	2	3	5	2	3	5	2	3	5	2	3	5	2	3	5	2	3	5	2	3	5
	>50%	3	4	5	3	4	5	3	4	5	3	4	5	3	4	5	3	4	5	3	4	5	3	4	5	3	4	5	3	4	5
Loss of camber - UR3	<10%	1	3	4	1	3	4	1	3	4	1	3	4	1	3	4	1	3	4	1	3	4	1	3	4	1	3	4	1	3	4
	10-50%	2	3	5	2	3	5	2	3	5	2	3	5	2	3	5	2	3	5	2	3	5	2	3	5	2	3	5	2	3	5
	>50%	3	4	5	3	4	5	3	4	5	3	4	5	3	4	5	3	4	5	3	4	5	3	4	5	3	4	5	3	4	5
Gravel thickness - UR4	<10%	1	3	5	1	3	5	1	3	5	1	3	5	1	3	5	1	3	5	1	3	5	1	3	5	1	3	5	1	3	5
	10-50%	2	3	5	2	3	5	2	3	5	2	3	5	2	3	5	2	3	5	2	3	5	2	3	5	2	3	5	2	3	5
	>50%	3	4	5	3	4	5	3	4	5	3	4	5	3	4	5	3	4	5	3	4	5	3	4	5	3	4	5	3	4	5
Erosion gullies - UR5	<10%	1	3	5	1	3	5	1	3	5	1	3	5	1	3	5	1	3	5	1	3	5	1	3	5	1	3	5	1	3	5
	10-50%	2	4	5	2	4	5	2	4	5	2	4	5	2	4	5	2	4	5	2	4	5	2	4	5	2	4	5	2	4	5
	>50%	3	5	5	3	5	5	3	5	5	3	5	5	3	5	5	3	5	5	3	5	5	3	5	5	3	5	5	3	5	5
Potholes - UR6	<5 (No./100m)	1	3	5	1	3	5	1	3	5	1	3	5	1	3	5	1	3	5	1	3	5	1	3	5	1	3	5	1	3	5
	5-15	2	4	5	2	4	5	2	4	5	2	4	5	2	4	5	2	4	5	2	4	5	2	4	5	2	4	5	2	4	5
	> 15	3	5	5	3	5	5	3	5	5	3	5	5	3	5	5	3	5	5	3	5	5	3	5	5	3	5	5	3	5	5
Clay - UR7	< 5%	1			1			1			1			1			1			1			1			1			1		
	5-50%	2			2			2			2			2			2			2			2			2			2		
	>50%	3			3			3			3			3			3			3			3			3			3		
....																															

Remarks:

OECD DC1	DAMAGE SHEET	DVI/UR 1
TYPE: RUTTING		CODE: RU

Rutting in both lanes
Extent 3
Severity 2
Class 5

DESCRIPTION:

Permanent longitudinal deformations following the line of vehicle wheel paths. In extreme cases the cross-section of the road shows a w-profile.

INSPECTION METHOD:

Severity:

Place a 2 m straightedge across the ruts and measure with a wedge the depth of the ruts in mm. Measure all ruts observed over the width of the road; representative value is the highest value measured.

Extent:

Percentage of sub-section length.

RATING CRITERIA:

Evaluation			Class				
	Extent	Severity			Severity		
					1	2	3
1	< 10 %	< 20 mm		1	1	3	5
2	10-50 %	20-50 mm	Extent	2	2	4	5
3	> 50 %	> 50 mm		3	3	5	5

OECD DC1	DAMAGE SHEET	DVI/UR 1

TYPE: RUTTING CODE: RU

PROBABLE CAUSES AND FACTORS AFFECTING THE DEVELOPMENT OF THE DAMAGE:

Traffic:

Resulting from wheel forces. Strongly influenced by traffic intensity, speed, loading or transverse distribution. The development of ruts is accelerated by heavy traffic and channelised traffic.

Climate:

Dry season:	Lateral displacement of non-cohesive materials.
Rainy season:	Liquifaction of surface material or stability loss of road embankment or supporting soils.

Materials:

Dry:	Increased risk for non-cohesive materials (mainly sandy).
Humid:	Increased risk for clayey materials and materials susceptible to water.

Others:

The presence of ruts may cause other damage to develop, such as longitudinal erosion gullies or potholes in weak spots in the surfacing.

REMARKS:

This damage has a bad effect on the safety of road users. Ruts prevent lateral movement of vehicles which accelerates rut development. In extreme cases the depth of ruts is such that the road becomes impassable.

POSSIBLE MAINTENANCE ACTIVITY:

1. No maintenance required;

2. Dragging of surface course;

3. Dry grading or dragging;

4. Wet grading or regravelling of surface course;

5. Heavy grading or rehabilitation of road.

OECD DC1	DAMAGE SHEET	DVI/UR 2
TYPE:	CORRUGATIONS	CODE: CO

Generalised corrugation.

Extent 3
Severity 3
Class 5

DESCRIPTION:

Shallow and evenly spaced ridges of surface material, transverse to road axis. These ridges extend over the entire width of the road; spacing varies between 500 and 1 000 mm depending on the average speed of vehicles.

INSPECTION METHOD:

Road condition survey:

Evident from within a vehicle either visually or by experiencing vibration.

Detailed visual inspection:

Measure by placing a 2 m straightedge across the top of the ridges; depth to be measured as well as distance between two successive ridges.

RATING CRITERIA:

Evaluation: Extent		Evaluation: Severity			Class: Severity 1	2	3
1	< 10 %	< 20 mm		1	1	3	4
2	10-50 %	20-50 mm	Extent	2	2	3	5
3	> 50 %	> 50 mm		3	3	4	5

OECD DC1	DAMAGE SHEET	DVI/UR 2
TYPE:	CORRUGATIONS	CODE: CO

PROBABLE CAUSES AND FACTORS AFFECTING THE DEVELOPMENT OF THE DAMAGE:

Traffic:

Lack of cohesion in the surface material and displacement by moving vehicles. The frequency (± 15 Hz) of the vibration of the suspended mass of vehicles combined with their speed determines the spacing of the ridges. The damage is occasioned by, and continues to develop through, existing irregularities in the surface.

Climate:

During the rainy season the ridge will also develop in the sub-base. The loss of course materials from the valleys will create weak spots leading to other damage.

The damage develops during the dry season when the material has little cohesion. The dryness and wind provide favourable conditions for the loss of fines.

Materials:

Materials most susceptible show little cohesion, a low plasticity index, contain particles larger than 5 mm and relatively few fines.

REMARKS:

This damage, one of the main sources of driver discomfort, is extremely harmful to vehicles. It is one of the main causes of increased vehicle operating costs.

POSSIBLE MAINTENANCE ACTIVITY:

1. No maintenance required;

2. Dragging of surface;

3. Dragging or light grading of surface;

4. Grading of surface;

5. Wet grading of surface.

TYPE: CAMBER/CROSSFALL CODE: CR

Loss of a crossfall,
water accumulating.

DESCRIPTION:

Distortion and deformation of the road cross-section. The ideal profile that should be maintained has a "roof" shape with transverse slopes of 3-4 % to permit the rapid flow of surface water off the road.

INSPECTION METHOD:

Severity:

The measurement can be performed rapidly with a straightedge equipped with a level tube and adapted to the required slope. The straightedge is placed transverse to the centreline; one end is placed on the centreline. The other end is raised by a height h (mm) to bring the straightedge to a level position, the gap between the underside of the straightedge and the road surface being a measure of the crossfall.

Extent:

Percentage of surface.

RATING CRITERIA:

Evaluation							Class		
	Extent		Severity (h)				Severity 1	2	3
1	< 10 %		> 50 mm			1	1	3	4
2	10-50 %		20-50 mm		Extent	2	2	3	5
3	> 50 %		< 20 mm			3	3	4	5

TYPE: CAMBER/CROSSFALL CODE: CR

PROBABLE CAUSES AND FACTORS AFFECTING THE DEVELOPMENT OF THE DAMAGE:

-- Wear by traffic;

-- Differential settlement;

-- Movement of materials.

REMARKS:

Defects in cross section reduce the removal of surface water thus weakening the road construction.

The formation of potholes and rutting will be accelerated.

POSSIBLE MAINTENANCE ACTIVITY:

1. Continue the recurrent maintenance;

2. Dry grading; shorten the interval of operations;

3. Dry grading;

4. Wet grading or regravelling of surface course;

5. Regravelling of surface course or rehabilitation of road.

OECD DC1	DAMAGE SHEET	DVI/UR 4

TYPE:	GRAVEL THICKNESS	CODE: GR

GR

Thickness between 15 and 20cm.

DESCRIPTION:

Thickness reduction of surface course.

Loss of surface material due to various causes including regrading and regravelling operations.

NOT RELEVANT FOR EARTH ROADS.

INSPECTION METHOD:

Severity is defined as the difference between the specified thickness and the measured thickness.

Road condition survey:

Monitor formation of ridges of material parallel to road axis. Extent expressed by % length affected.

Detailed visual inspection:

Measure every 500 m remaining thickness of surface course by digging hole to sub-base level (see picture above). Fill hole after measurement by replacing excavated material.

RATING CRITERIA:

Evaluation				Class		
Extent	Severity			Severity		
				1	2	3
1 ‹ 10 %	‹ 10 mm		1	1	3	5
2 10-50 %	10-50 mm	Extent	2	2	3	5
3 › 50 %	› 50 mm		3	3	4	5

OECD DC1	DAMAGE SHEET	DIV/UR 4

TYPE:	GRAVEL/THICKNESS	CODE: GR

PROBABLE CAUSES AND FACTORS AFFECTING THE DEVELOPMENT OF THE DAMAGE:

Traffic:

Is one of the main causes of loss of material. An ADT of about 140 vehicles leads to a loss of material of 10 to 30 mm per annum.

Climate:

Losses occur during all seasons, however rain will increase the loss of material. In the dry season this damage will also result in dust generated by traffic.

Materials:

Strongly dependent on type of material used for construction.

Others:

Combined action of all factors including maintenance, affects the pavement (or surface course). Development of damage depends on region, climate, materials used for road construction, traffic intensity, and topography.

REMARKS:

Road users safety is endangered by dust generated by traffic; dust reduces the visibility and is also a nuisance for people living adjacent to the road.

POSSIBLE MAINTENANCE ACTIVITY:

1. No maintenance required;
2. Dragging of surface;
3. Dry grading of road surface;
4. Regravelling of surface course or wet grading;
5. Reconstruction of surface course.

OECD DC1	DAMAGE SHEET	DIV/UR 5

TYPE: EROSION GULLIES CODE: EG

Longitudinal erosion gullies.

Extent 2
Severity 3
Class 5

DESCRIPTION:

The aspect is dependent upon the mode of development and the location.

Three main types are distinguished:
1) Scouring of side drains, following the slope of the formation level and affecting a section with steep slopes and bad compaction.
2) Longitudinal gulleys parallel to the road axis in the surface.
3) Gulleys transverse to the road axis cutting the road over its entire width.

INSPECTION METHOD:

Road condition survey:

Estimate and record the location of dangerous gullies.

Detailed visual inspection:

Place a 2 m straightedge across the gullies and measure their depth in mm with a tape or rule.

RATING CRITERIA:

Evaluation: Extent	Evaluation: Severity		Class: Severity 1	2	3
1 < 10 %	< 20 mm	Extent 1	1	3	5
2 10-50 %	20-50 mm	Extent 2	2	4	5
3 > 50 %	> 50 mm	Extent 3	3	5	5

OECD DC1	DAMAGE SHEET	DVI/UR 5

TYPE: EROSION GULLIES CODE: EG

PROBABLE CAUSES AND FACTORS AFFECTING THE DEVELOPMENT OF THE DAMAGE:

Traffic:

Traffic may create damage leading to erosion. Ruts develop into longitudinal gullies; corrugations may create perpendicular gullies.

Climate:

The severity of the erosion depends on the quantity and the velocity of the surface water to be removed.

Materials:

Badly compacted materials and materials likely to be affected by water.

Others:

Drainage and the quality of the surface courses are important. Damage occurs mainly in places where slopes are steep.

1. Side gullies	:	Resulting from normal discharge of water draining along the slope of the terrain through soft materials.
2. Longitudinal gullies	:	Poor cross-section; presence of ruts; improper maintenance of shoulders and deficiencies in drainage capacity.
3. Transverse gullies	:	Flooding of side drains; drainage of water along the natural slope of the terrain.

REMARKS:

Erosion develops quickly and can lead to total severance of the road. Deep gullies endanger the safety of road users.

POSSIBLE MAINTENANCE ACTIVITY:

1-2. Monitor drainage and cross-section of the road.

3. If prominent, grading or any other maintenance technique considered suitable for the damage leading to erosion gullies.

4-5. Extensive damage: regravelling or reconstruction of the road; if damage is localised fill gulley and restore road profile.

TYPE: POTHOLES CODE: PO

Pothole on dry roadway.

Extent 2
Severity 3
Class 5

DESCRIPTION:

In general bowl shaped holes caused by the loss of surface material.

INSPECTION METHOD:

Road condition survey:

Visual recording and estimation of extent; driving conditions; recording of deep potholes.

Detailed visual inspection:

Extent : Evaluation/assessment of number of potholes per 100 m of road length.

Severity : Place a straightedge over pothole and measure the depth of the hole.

RATING CRITERIA:

Evaluation: Extent No. per 100m	Evaluation: Severity			Class Severity 1	2	3
1 < 5	< 20 mm		1	1	3	5
2 5-15	20-40 mm	Extent	2	2	4	5
3 > 15	> 40 mm		3	3	5	5

OECD DC1	DAMAGE SHEET	DVI/UR 6

TYPE: POTHOLES CODE: PO

PROBABLE CAUSES AND FACTORS AFFECTING THE DEVELOPMENT OF THE DAMAGE:

Traffic:

Potholes develop in those areas where the subgrade is uneven or results from other damage (corrugations, rutting, erosion, gullies, etc.). Increased traffic will accelerate the development of potholes.

Climate:

Potholes start to develop in the wet season; water remains in them and soak the surface material, making it vulnerable to further damage. Once present, potholes will grow deeper and wider.

Materials:

Impermeable clayey soils prevent proper drainage of surface water and permit weak spots to develop into potholes.

Others:

This damage often results from neglected problems such as corrugations, rutting or drainage. Corrugations and rutting may cause the development of a series of potholes.

REMARKS:

Once potholes are present, they are to be repaired immediately since they worsen quickly. Potholes have an adverse effect on the safety and comfort of the road user.

POSSIBLE MAINTENANCE ACTIVITY:

1. No maintenance required.

2. Fill potholes.

3. Fill potholes.

4. Grading or regravelling of surface.

5. Wet grading or regravelling of surface.

OECD DC1	DAMAGE SHEET	DVI/UR 7

TYPE:	CLAY	CODE: CY

Road surface extensively covered by clay.

Extent 3

DESCRIPTION:

Presence of clayey materials or fines in thin layers on top of the surface course. Shiny appearance during rainy season. Smooth appearance with cracks during dry weather.

INSPECTION METHOD:

Extent:

Estimation of percentage of surface affected.

RATING CRITERIA:

1. < 5 %
2. 5-50 %
3. > 50 %

OECD DC1	DAMAGE SHEET	DVI/UR 7

TYPE:	CLAY	CODE: CY

PROBABLE CAUSES AND FACTORS AFFECTING THE DEVELOPMENT OF THE DAMAGE:

Loss of (gravel) base course.

Contamination of surface course by fines from lower courses.

During dry weather this type of damage may lead to other faults, similar to those appearing in paved roads, e.g. cracks, potholes, etc.

REMARKS:

Wet season: Surface can be dangerously slippery.

Dry season: Uneven appearance of surface can be a danger.

POSSIBLE MAINTENANCE ACTIVITY:

1-2. No maintenance required.

3. Regravelling of surface.

SECTION 4

DETAILED VISUAL INSPECTION OF PAVED ROADS

Type	No. of sheet	Code	Function affected	Rating levels
Rutting without cracks	DVI/PR 1a	RU	Structure/ Roughness/ Drainage	5
Rutting with cracks	DVI/PR 1b	RU/C	Structure/ Roughness/ Drainage	5
Corrugations	DVI/PR 2	CO	Carriageway	5
Depressions	DVI/PR 3	DP	Drainage	5
Transverse cracks	DVI/PR 4a	CT	Structure	5
Longitudinal cracks	DVI/PR 4b	CL	Structure	5
Alligator cracks	DVI/PR 4c	CA	Structure	5
Holes	DVI/PR 5	HO	Structure	5
Edge distress	DVI/PR 6	ED	Structure	5
Stripping/fretting/ ravelling	DVI/PR 7a	ST	Structure	5
Stripping/fretting/ ravelling of surface	DVI/PR 7b	ST/S	Structure	5
Bleeding	DVI/PR 8	BL	Safety	3

Figure 7. **DETAILED VISUAL INSPECTION FORM FOR PAVED ROADS ("FORM VII")**

DETAILED VISUAL INSPECTION (DVI)					**PAVED ROADS**
OFFICE:	DISTRICT:		ROAD DENOMINATION/ CLASS:		Main Section No.
Name of inspector:			FROM I.P.:		Km:
Date:	Weather: o Clear o Rainy	Carr.way: o Dry o Drying o Wet	TO I.P.:		Km:
			Pavement type:	Pavement width: m	Section length: Km

SUB-SECTION		◯						◯						◯						◯						◯					
Roadside		L			R			L			R			L			R			L			R			L			R		
Damage type - Cat	E \ S	L	M	S	L	M	S	L	M	S	L	M	S	L	M	S	L	M	S	L	M	S	L	M	S	L	M	S	L	M	S
Rutting - PR1	<10%	1	3	5	1	3	5	1	3	5	1	3	5	1	3	5	1	3	5	1	3	5	1	3	5	1	3	5	1	3	5
	10-50%	2	4	5	2	4	5	2	4	5	2	4	5	2	4	5	2	4	5	2	4	5	2	4	5	2	4	5	2	4	5
	>50%	3	5	5	3	5	5	3	5	5	3	5	5	3	5	5	3	5	5	3	5	5	3	5	5	3	5	5	3	5	5
Corrugations - PR2	<10%	1	3	4	1	3	4	1	3	4	1	3	4	1	3	4	1	3	4	1	3	4	1	3	4	1	3	4	1	3	4
	10-50%	2	3	5	2	3	5	2	3	5	2	3	5	2	3	5	2	3	5	2	3	5	2	3	5	2	3	5	2	3	5
	>50%	3	4	5	3	4	5	3	4	5	3	4	5	3	4	5	3	4	5	3	4	5	3	4	5	3	4	5	3	4	5
Depressions - PR3	<10%	1	3	4	1	3	4	1	3	4	1	3	4	1	3	4	1	3	4	1	3	4	1	3	4	1	3	4	1	3	4
	10-50%	2	4	5	2	4	5	2	4	5	2	4	5	2	4	5	2	4	5	2	4	5	2	4	5	2	4	5	2	4	5
	>50%	3	5	5	3	5	5	3	5	5	3	5	5	3	5	5	3	5	5	3	5	5	3	5	5	3	5	5	3	5	5
Transverse cracking - PR4a	<2 (No./100m)	1	3	4	1	3	4	1	3	4	1	3	4	1	3	4	1	3	4	1	3	4	1	3	4	1	3	4	1	3	4
	2-15	2	3	5	2	3	5	2	3	5	2	3	5	2	3	5	2	3	5	2	3	5	2	3	5	2	3	5	2	3	5
	>15	3	4	5	3	4	5	3	4	5	3	4	5	3	4	5	3	4	5	3	4	5	3	4	5	3	4	5	3	4	5
Longitudinal cracking - PR4b	<10%	1	3	5	1	3	5	1	3	5	1	3	5	1	3	5	1	3	5	1	3	5	1	3	5	1	3	5	1	3	5
	10-50%	2	4	5	2	4	5	2	4	5	2	4	5	2	4	5	2	4	5	2	4	5	2	4	5	2	4	5	2	4	5
	>50%	3	5	5	3	5	5	3	5	5	3	5	5	3	5	5	3	5	5	3	5	5	3	5	5	3	5	5	3	5	5
Alligator cracking - PR4c	<10%	1	4	5	1	4	5	1	4	5	1	4	5	1	4	5	1	4	5	1	4	5	1	4	5	1	4	5	1	4	5
	10-50%	2	5	5	2	5	5	2	5	5	2	5	5	2	5	5	2	5	5	2	5	5	2	5	5	2	5	5	2	5	5
	>50%	3	5	5	3	5	5	3	5	5	3	5	5	3	5	5	3	5	5	3	5	5	3	5	5	3	5	5	3	5	5
Holes - PR5	<5 (No./100m)	1	3	5	1	3	5	1	3	5	1	3	5	1	3	5	1	3	5	1	3	5	1	3	5	1	3	5	1	3	5
	5-15	3	4	5	3	4	5	3	4	5	3	4	5	3	4	5	3	4	5	3	4	5	3	4	5	3	4	5	3	4	5
	> 15	4	5	5	4	5	5	4	5	5	4	5	5	4	5	5	4	5	5	4	5	5	4	5	5	4	5	5	4	5	5
Edge distress - PR6	<10%	1	3	4	1	3	4	1	3	4	1	3	4	1	3	4	1	3	4	1	3	4	1	3	4	1	3	4	1	3	4
	10-50%	2	3	5	2	3	5	2	3	5	2	3	5	2	3	5	2	3	5	2	3	5	2	3	5	2	3	5	2	3	5
	>50%	3	4	5	3	4	5	3	4	5	3	4	5	3	4	5	3	4	5	3	4	5	3	4	5	3	4	5	3	4	5
Stripping/ Fretting/ Ravelling - PR7a	<10%	1	2	4	1	2	4	1	2	4	1	2	4	1	2	4	1	2	4	1	2	4	1	2	4	1	2	4	1	2	4
	10-50%	1	3	5	1	3	5	1	3	5	1	3	5	1	3	5	1	3	5	1	3	5	1	3	5	1	3	5	1	3	5
	>50%	2	5	5	2	5	5	2	5	5	2	5	5	2	5	5	2	5	5	2	5	5	2	5	5	2	5	5	2	5	5
Stripping/Fretting/ Ravelling of surface - PR7b	<15%	1	2	3	1	2	3	1	2	3	1	2	3	1	2	3	1	2	3	1	2	3	1	2	3	1	2	3	1	2	3
	15-30%	2	4	5	2	4	5	2	4	5	2	4	5	2	4	5	2	4	5	2	4	5	2	4	5	2	4	5	2	4	5
	>30%	3	5	5	3	5	5	3	5	5	3	5	5	3	5	5	3	5	5	3	5	5	3	5	5	3	5	5	3	5	5
Bleeding - PR8	< 5%				1			1			1			1			1			1			1			1			1		
	5-50%				2			2			2			2			2			2			2			2			2		
	>50%				3			3			3			3			3			3			3			3			3		

Remarks: ______________________________

OECD DC1	DAMAGE SHEET	DVI/PR 1a

TYPE: RUTTING WITHOUT CRACKS CODE: RU

Rutting without cracks due to
heavy trucks with twin-wheels.

Extent 3
Severity 2
Class 5

DESCRIPTION:

More likely to occur in pavements with thick bituminous layers. Permanent depressions in the road surface affecting the road cross-section in the wheel paths. Generally over long distances. In some instances, ruts will be bordered by excess material displaced by the ruts.

INSPECTION METHOD:

<u>Severity</u>:

-- Place 2 m straightedge over the wheel paths and measure the depth of the ruts;
-- Take several measurements over the section or sub-section length;
-- Take highest value measured as representative measure;
-- Distance between measurements: 500 m in section, 100 m in sub-section.

<u>Extent</u>:

Percentage of length.

RATING CRITERIA:

Evaluation					Class		
Extent		Severity			Severity		
					1	2	3
1	< 10 %	< 20 mm		1	1	3	5
2	10-50 %	20-40 mm	Extent	2	2	4	5
3	> 50 %	> 40 mm		3	3	5	5

TYPE: RUTTING WITHOUT CRACKS CODE: RU

PROBABLE CAUSES AND FACTORS AFFECTING THE DEVELOPMENT OF THE DAMAGE:

Accumulation of irreversible deformation and/or lateral material displacement within the pavement layers.

Traffic:

Influence of axle load and traffic intensity is important. Low speeds are the most damaging. Narrow roads present higher risk as vehicles tend to follow the same wheel path. Double ruts can then be observed resulting from the action of dual tyres.

Climate:

High temperatures increase the risks of deformation in the bituminous layers.

Bituminous materials:

Wrong mix design; too much/too little bitumen; too many fines; poor interlock between aggregates; incorrect bitumen type; poor compaction.

REMARKS:

Damage will reduce lateral movement of vehicles thus increasing the development of existing ruts.

Presence of water in the ruts will endanger traffic through an increased risk of "aquaplaning".

POSSIBLE MAINTENANCE ACTIVITY:

Replace bituminous layers by materials of better composition.

TYPE: **RUTTING WITH CRACKS** CODE: RU/C

Wheeltrack rutting and cracking.

Extent 3
Severity 2
Class 5

DESCRIPTION:

Concerns flexible pavements with thin layers of bituminous material. Permanent depressions in the road surface accompanied by cracks. Affects the road cross-section over the full length of the sub-section. This damage occurs especially in the wheel tracks and also along the edge of the road.

INSPECTION METHOD:

Severity:

-- Place a 2 m straightedge across the wheel paths and measure the depth of the ruts;
-- Take several measurements over the section or sub-section length;
-- For the representative value take the highest value measured;
-- Distance between measurements: 500 m in section, 100 m in sub-section.

Extent:

-- Percentage of length.

RATING CRITERIA:

-- To be rated in accordance with rating criteria for rutting and cracking;
-- In this case rutting and cracking are both to be recorded on the DVI-form.

OECD DC1 DAMAGE SHEET DVI/PR 1b

TYPE: RUTTING WITH CRACKS CODE: RU/C

PROBABLE CAUSES AND FACTORS AFFECTING THE DEVELOPMENT OF THE DAMAGE:

Inadequate pavement design.
Rigidity and thickness of top layers not appropriate to traffic and bearing capacity conditions.

Traffic:

Important influence of axle load and traffic intensity. Narrow roads present higher risk as vehicles tend to follow the same wheel paths.

Climate:

During rainy season increased moisture content in soil and foundation layers enhance the risk.
Proper drainage has to be maintained.

Materials:

High risks with clayey soils, lateritic soils, contaminated base courses or base courses containing too much fines.

REMARKS:

Damage will reduce lateral movement of vehicles thus increasing the development of existing ruts.

Presence of water in the ruts will endanger traffic through an increased risk of "aquaplaning".

POSSIBLE MAINTENANCE ACTIVITY:

1.
2-3. Improve foundation by improving drainage; seal cracks.
4-5. Improve foundation, reinforce pavement structure.

TYPE: CORRUGATIONS CODE: CO

Transverse undulations
at intersection.

DESCRIPTION:

Shallow and evenly spaced ridges of surface material, perpendicular to the longitudinal road axis. Spacing of the ridges is often about 1 m; this type of damage is often localised.

INSPECTION METHOD:

Road condition survey:

Noticeable from within a vehicle either visually or by experiencing vibrations.

Detailed visual inspection:

Severity: Measurements with straightedge placed on the tops of the corrugations and determination of depth (h) with measuring wedge.

Extent: Affected area as a percentage of the total area.

RATING CRITERIA:

Evaluation			Class			
Extent	Severity (h)			Severity		
				1	2	3
1 < 10 %	< 20 mm		1	1	3	4
2 10-50 %	20-40 mm	Extent	2	2	3	5
3 > 50 %	> 40 mm		3	3	4	5

TYPE: CORRUGATIONS CODE: CO

PROBABLE CAUSES AND FACTORS AFFECTING THE DEVELOPMENT OF THE DAMAGE:

Corrugation occurs if one or more pavement layers start to flow, thus permitting movement of materials into the corrugated surface.

Traffic:

Tangential forces from wheels are the main cause of this type of damage, particularly where vehicles frequently stop or accelerate.
Depressions may initiate the development of corrugations.

Climate:

High temperature increases the risk of such damage.

Wearing course design

Lack of adhesion between base and wearing course.

Materials:

Lack of stiffness in one of the layers due to improper construction, execution or design.

REMARKS:

POSSIBLE MAINTENANCE ACTIVITY:

1-2-3. No action required;
4-5. Replace corrugated layers; choose mixtures with a better composition and quality.

TYPE: DEPRESSIONS CODE: DP

Large depression with alligator cracking.

Extent 1
Severity 3
Class 4

DESCRIPTION:

Vertical settlement of the road surface, usually localised and circular in shape. Often the surface material is cracked.

INSPECTION METHOD:

Severity:

Place a 2 m straightedge over the affected area and measure the depth of the depression with a measuring wedge.

Extent:

Affected area as a percentage of the total area.

RATING CRITERIA:

Evaluation				Class		
Extent	Severity			Severity		
				1	2	3
1 < 10 %	< 20 mm		1	1	3	4
2 10-50 %	20-40 mm	Extent	2	2	4	5
3 > 50 %	> 40 mm		3	3	5	5

TYPE: DEPRESSIONS CODE: DP

PROBABLE CAUSES AND FACTORS AFFECTING THE DEVELOPMENT OF THE DAMAGE:

Settlement of lower layers and local loss of bearing capacity due to excessive moisture content within the pavement.

Traffic:

Heavy traffic is a major factor and worsens this damage rapidly.

Drainage:

Bad surface drainage	:	due to localised low spots. Run-off of surface water is not adequate. The water then penetrates into the foundation layers.
Bad lateral drainage	:	Water is ponding in sidedrains, raising the water table in the subgrade.

Climate:

Increased risks during rainy season.

Materials:

Clayey soils and clay contaminated subbase and base course materials.

REMARKS:

Once present, this type of damage worsens rapidly.

POSSIBLE MAINTENANCE ACTIVITY:

1. No action required.
2. Maintain drains. Reinstate the surface course.
3. Maintain drains. Reinstate the surface course.
4. Maintain drains. Reinstate the surface course.
5. Maintain drains. Remove the material in the affected area; replace by new base or subbase material and compact properly; reinstate the surface course.

TYPE: TRANSVERSE CRACKS CODE: CT

Crack due to shrinking of base.

Severity 2

DESCRIPTION:

Rupture over the full width of the road.

INSPECTION METHOD:

<u>Severity</u>:

Measurement of width of crack in mm.

<u>Extent</u>:

Count all transverse cracks surpassing 1.5 m in length over a section of 100 m.

RATING CRITERIA:

Evaluation: Extent No.	Extent per 100 m	Severity width			Class Severity 1	2	3
1	< 2	< 2 mm		1	1	3	4
2	2-15	2-10 mm	Extent	2	2	3	5
3	> 15	> 10 mm		3	3	4	5

TYPE: TRANSVERSE CRACKS CODE: CT

PROBABLE CAUSES AND FACTORS AFFECTING THE DEVELOPMENT OF THE DAMAGE:

-- Inadequate joint construction.
-- Lack of compaction of a road section.
-- Reflection cracking from foundation layers.
-- Structural discontinuity of pavement structure.

Climate:

During rainy season, water penetrates through the cracks to the underlaying layers, thus weakening the sub-soil and leading to failure.
Due to temperature variations, movement of the pavement layers will cause stresses at joints and in the foundation layers which will result in reflection cracking.

Materials:

Cement stabilized foundation materials (shrinkage cracks during drying process).
Lime stabilized foundation materials.

REMARKS:

This type of damage may lead to other types when water infiltrates through the cracks into the sub-soil (settlement, alligator cracking, etc.).

POSSIBLE MAINTENANCE ACTIVITY:

1-2-3. Seal cracks.

4-5. Cut away wearing course and apply a new layer.

TYPE: LONGITUDINAL CRACKS CODE: CL

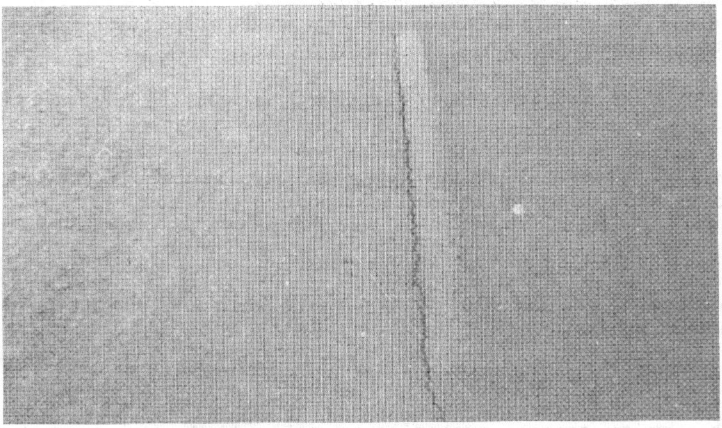

Longitudinal cracking
on the centre line.

DESCRIPTION:

Crack or rupture parallel to the road axis (central or edge).

INSPECTION METHOD:

Severity:

Measurement of width of crack in mm.

Extent:

Evaluate percentage of section length affected by this type of damage.

RATING CRITERIA:

Evaluation					Class		
Extent		Severity			Severity		
					1	2	3
1	< 10 %	< 2 mm		1	1	3	5
2	10-50 %	2-10 mm	Extent	2	2	4	5
3	> 50 %	> 10 mm		3	3	5	5

TYPE: LONGIDUDINAL CRACKS CODE: CL

PROBABLE CAUSES AND FACTORS AFFECTING THE DEVELOPMENT OF THE DAMAGE:

Different causes may be distinguished:

-- Pavement fatigue (see also rutting accompanied by cracks); problems due to design;

-- Opening of a longitudinal joint in the wearing course (construction fault);

-- Reflection cracks where pavement has been widened; (uneven settlement of foundation or movement of longitudinal joints).

Traffic:

The influence of traffic will be important when the damage is concentrated in the wheel tracks.

Climate:

During the rainy season, weakening of sub-soil due to water infiltration. The variation in temperature will induce movement of materials and loss of interlock and adhesion of bitumen.

Materials:

Wearing course too fragile; bad mix design; low bearing capacity of foundation layers and soil.

REMARKS:

POSSIBLE MAINTENANCE ACTIVITY:

1-2-3. Seal cracks;

4-5. Cut away wearing course and construct a new layer or overlay.

TYPE: ALLIGATOR CRACKING CODE: CA

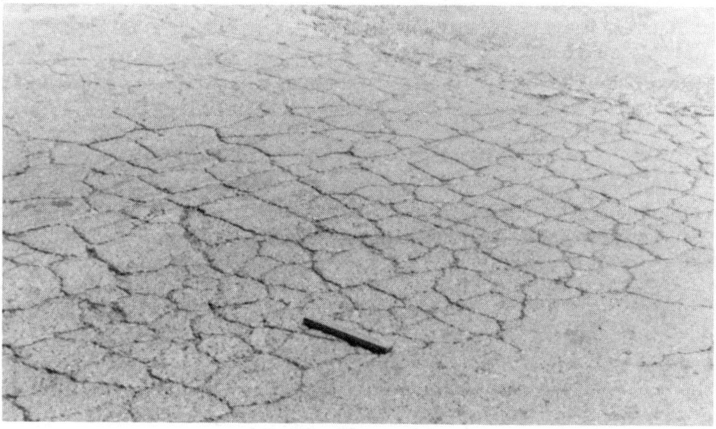

Interconnected cracking.

DESCRIPTION:

Network of cracks in all directions, linked together; localised in the wheel tracks or over the full width of the road.

INSPECTION METHOD:

Severity: Measurement of width of crack in mm.

Extent: Detailed visual inspection: evaluate percentage of area affected.

RATING CRITERIA:

Evaluation					Class		
Extent		Severity			Severity		
					1	2	3
1	< 10 %	< 2 mm		1	1	4	5
2	10-50 %	2-10 mm	Extent	2	2	5	5
3	> 50 %	> 10 mm		3	3	5	5

TYPE: ALLIGATOR CRACKING CODE: CA

PROBABLE CAUSES AND FACTORS AFFECTING THE DEVELOPMENT OF THE DAMAGE:

Design:

-- Inadequate for traffic load; insufficient thickness of surface layers. Pavement is at the end of its life; low bearing capacity of soil.

Traffic:

The damage appears in the first instance in the wheel tracks where the effects of traffic are greatest. If the base and subbase course are weak, this damage will be accompanied by deformation (see under "rutting with cracks").

Climate:

During the rainy season, weakening of sub-soil due to water infiltration. When the drainage is insuffiicient the damage will occur in the first instance in the outer wheel path of the road.

Materials:

Top layer:	Too stiff relative to its thickness. Premature ageing of material.
Lower layers:	Contaminated by clayey soils; clay or laterite soils.

REMARKS:

Ultimately alligator cracking will result in the formation of potholes.

POSSIBLE MAINTENANCE ACTIVITY:

1. None;

2-3. Superficial treatment, sealing;

4-5. Cut away and construct a new layer.

OECD DC1	DAMAGE SHEET	DVI/PR 5

TYPE: HOLES CODE: HO

Holes in wheeltracks.

Extent 1
Severity 3
Class 5

DESCRIPTION:

In general round holes caused by the loss of material in the top layer.

INSPECTION METHOD:

Road condition survey:

Visual recording and estimation of extent; driving conditions; localised deep holes.

Detailed visual inspection:

Severity: Place a straightedge over pothole and measure the depth of the hole with a tape or rule.

Extent: Counting the number of potholes per 100 m of road length.

RATING CRITERIA:

Evaluation					Class		
Extent		Severity			Severity		
No. per 100 m					1	2	3
1	< 5	< 20 mm		1	1	3	5
2	5-15	20-40 mm	Extent	2	3	4	5
3	> 15	> 40 mm		3	4	5	5

OECD DC1	DAMAGE SHEET	DVI/PR 5

TYPE: HOLES CODE: HO

PROBABLE CAUSES AND FACTORS AFFECTING THE DEVELOPMENT OF THE DAMAGE:

Traffic:

Holes develop in those places where the subgrade is deformed or as a result of other damage (alligator cracking, settlements, etc.); increased traffic will accelerate the development of holes.

Climate:

Holes start to develop in the wet season; water remains in the depressions and soaks the foundation making it vulnerable to traffic. Once established potholes will continue to develop during all seasons.

Materials:

Wearing course material fragile or premature ageing of binder; clayey soils or contaminated foundations prevent proper drainage and encourage the development of holes.

Others:

This type of damage often results from neglected maintenance (surface treatment of road with alligator cracking).

REMARKS:

Once potholes are present, they are to be rapaired immediately since they worsen quickly. Potholes have an adverse effect on the safety and comfort of the road user.

POSSIBLE MAINTENANCE ACTIVITY:

1. No maintenance required;

2. Preventive maintenance of cracks (sealing);

3. Cut away bad material to obtain a patch with straightedges; fill with bituminous material and compact;

4-5. Cut away wearing course and apply a new bituminous layer.

Deep holes endangering traffic must be repaired immediately.

OECD DC1	DAMAGE SHEET	DVI/PR 6

TYPE: EDGE DISTRESS CODE: ED

Edge distress with shoulder deformation.

Extent 3
Severity 3
Class 5

DESCRIPTION:

Cracking and disintegration of the edges of the bituminous pavement.

INSPECTION METHOD:

<u>Severity</u>: Measure level differences.

<u>Extent</u>: Evaluate percentage of affected length.

RATING CRITERIA:

Evaluation: Extent	Evaluation: Severity			Class: Severity 1	Class: Severity 2	Class: Severity 3
1 < 10 %	< 10 mm		1	1	3	4
2 10-50 %	10-25 mm	Extent	2	2	3	5
3 > 50 %	> 25 mm		3	3	4	5

OECD DC1	DAMAGE SHEET	DVI/PR 6

TYPE: EDGE DISTRESS CODE: ED

PROBABLE CAUSES AND FACTORS AFFECTING THE DEVELOPMENT OF THE DAMAGE:

Traffic:

Due to narrow road width, vehicles are forced to use the edge of the road or the shoulder; settlement of shoulder due to parking of vehicles.

Construction:

Lack of support of shoulder; shoulder unstable and too low.

Materials:

Lack of adhesion in the surfacing ; lack of stiffness.

Climate:

Erosion will reduce shoulder level and progressively affect the surfacing.

REMARKS:

This type of damage tends to worsen rapidly and may affect the safety of traffic.

POSSIBLE MAINTENANCE ACTIVITY:

1. No maintenance required;
2. Repair areas affected;
3. Reconstruct shoulder and edge of pavement;
4. Reconstruct shoulder and edge of pavement;
5. Reconstruct completely.

TYPE: STRIPPING/FRETTING/RAVELLING CODE: ST

Localised damage.

Extent 1
Severity 2
Class 2

DESCRIPTION:

Loss of areas of the surfacing.

INSPECTION METHOD:

Severity: Measure thickness of surfacing course affected in mm.

Extent: Evaluate percentage of area affected.

RATING CRITERIA:

Evaluation: Extent		Evaluation: Severity			Class: Severity 1	2	3
1	< 10 %	< 10 mm		1	1	2	4
2	10-50 %	10-25 mm	Extent	2	1	3	5
3	> 50 %	> 25 mm		3	2	5	5

TYPE: STRIPPING/FRETTING/RAVELLING CODE: ST

PROBABLE CAUSES AND FACTORS AFFECTING THE DEVELOPMENT OF THE DAMAGE:

Climate:

In the case of insufficient adhesion, infiltrating water (through cracks or porous surface layer will accumulate between the pavement layers and weaken the pavement structure.

Materials:

Insufficient thickness or stability of wearing course; lack of bond with lower layer.

REMARKS:

Lack of maintenance of this type of damage will result in the formation of potholes.

POSSIBLE MAINTENANCE ACTIVITY:

1. None;

2. Localised repairs; surface treatment or sealing;

4-5. Cut away wearing course and reconstruct.

TYPE: STRIPPING/FRETTING/RAVELLING OF SURFACE CODE: ST/S

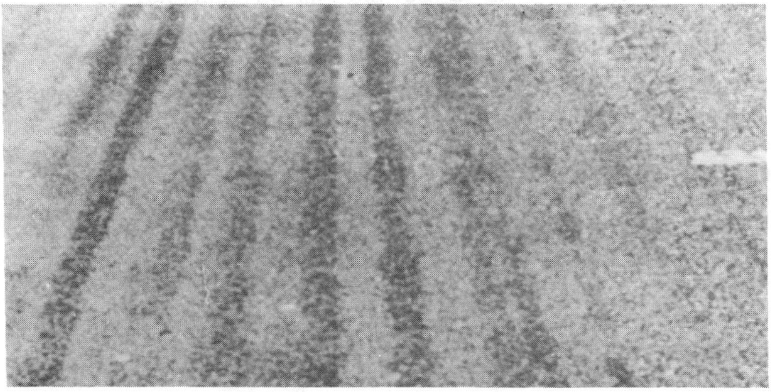

Longitudinal removal.

Extent 3

DESCRIPTION:

ONLY SURFACE DRESSINGS ARE CONCERNED

Removal of aggregates in longitudinal strips.

Pavement binder appears alternatively in thin and thick lines on the surface, parallel to road axis.

Removal of aggregates of a large surface area.

INSPECTION METHOD:

Severity: Evaluate percentage of a unit area (1 sq. m) affected
Extent: Evaluate percentage of the subsection affected.

RATING CRITERIA:

Evaluation					Class		
Extent		Severity			Severity		
					1	2	3
1	‹ 15 %	‹ 10 %		1	1	2	3
2	15-30 %	10-50 %	Extent	2	2	4	5
3	› 30 %	› 50 %		3	3	5	5

OECD DC1	DAMAGE SHEET	DVI/PR 7b

TYPE: STRIPPING/FRETTING/RAVELLING OF SURFACE — CODE: ST/S

PROBABLE CAUSES AND FACTORS AFFECTING THE DEVELOPMENT OF THE DAMAGE:

During laying:

Bad spreading of binder (blocked, badly adjusted or aimed nozzles).
Bad adjustment in height of the spreading bar;
Bad binder/aggregate bond;
Bad aggregate gradation of the surface mix;
Bad spreading of aggregates.

REMARKS:

POSSIBLE MAINTENANCE ACTIVITY:

-- Renew treatment with better control of laying operation.

TYPE: BLEEDING CODE: BL

Bleeding in wheeltracks.

Extent 3 right side
Extent 2 left side

DESCRIPTION:

Localised accumulation of bitumen at road surface, making the road appear black and shiny.

INSPECTION METHOD:

Extent: Evaluate percentage of surface area affected.

RATING CRITERIA:

Extent

1 < 5 %
2 5-50 %
3 > 50 %

OECD DC1	DAMAGE SHEET	DVI/PR 8
TYPE: BLEEDING		CODE: BL

PROBABLE CAUSES AND FACTORS AFFECTING THE DEVELOPMENT OF THE DAMAGE:

Climate:

High temperature softens the binder and makes it more susceptible to damage (binder too soft or too susceptible to high temperatures).

Materials and construction execution

Excess bitumen in surface course; damage will occur generally in wheel paths.

Upward movement of bitumen from underlying tack coat or previous surface layer.

REMARKS:

Bleeding makes the road surface slippery.

POSSIBLE MAINTENANCE ACTIVITY:

1. No actions required.
2. Spread chippings or sand on the affected area.
3. Remove the affected layer and replace.

Figure 8. **SUMMARY OF DETAILED VISUAL INSPECTION RESULTS: AVERAGE CONDITION VALUE ("FORM VIII")**

DETAILED VISUAL INSPECTION (DVI)

OFFICE:	DISTRICT:	PAGE
Established by:		
Date:		

DESIGNATION OF ROAD SECTION								
Road		Section / Sub-section No.	Identification Points (IP)			Section length	Carriageway width	AADT
Class	No.		I.P.	Designation	Km	Km	m	Veh/day
1	2	3	4	5	6	7	8	9
			From		From			
			To		To			
			From		From			
			To		To			
			From		From			
			To		To			
			From		From			
			To		To			
			From		From			
			To		To			
			From		From			
			To		To			
			From		From			
			To		To			
			From		From			
			To		To			
			From		From			
			To		To			
			From		From			
			To		To			

	Visual inspection data																		Avg.cond.value	Instrumental survey data		
	Unpaved roads								Paved roads													
Roadside	Rutting	Corrugations	Loss of camber	Gravel thickness	Erosion gullies	Potholes	Clay		Rutting	Corrugations	Depressions	Tr/Long cracking	Alligator crack.	Holes	Edge distress	Stripping	Bleeding		IRI	Deflection	Skid resistance
11	12	13	14	15	16	17	18	19	20	21	22	23	24	25	26	27	28	29	30	31	32
L																					
R																					
L																					
R																					
L																					
R																					
L																					
R																					
L																					
R																					
L																					
R																					
L																					
R																					
L																					
R																					
L																					
R																					
L																					
R																					

LIST OF GROUP MEMBERS

Chairman: Mr. Geir REFSDAL

Australia	Dr. J. METCALF Australian Road Research Board (ARRB) P.O. Box 156 (Bag 4) Nunawading, Victoria 3131
Belgium	Dr. Louis FRANCKEN Chercheur au service "Conception et structures" Centre de Recherches Routières Boulevard de la Woluwe 42 B-1200 Brussels
	Dr. Michel GORSKI Chercheur - Service Etudes Générales Centre de Recherches Routières Boulevard de la Woluwe 42 B-1200 Brussels
	Mr. Jean LECLERCQ Chargé de cours à la Faculté des Sciences agronomiques de l'Etat Service "Résistance des matériaux et construction du génie rural" B-5800 Gembloux
Denmark	Mr. Jan JANSEN National Road Laboratory Elisagaardsvej 5 DK-4000 Roskilde
Finland	Mr. Lars BJÖRKSTEN Roads and Waterways Administration (RWA) PL 33 SF-00521 Helsinki
France	Mr. C. PEYRONNE SETRA 46 Avenue Aristide Briand 92223 Bagneux

Mr. G. CAROFF
SCETAUROUTE
Technique et Développement
2 rue Stephenson
78181 St. Quentin-Yvelines Cédex

<u>Germany</u>

Mr. H. NEUBAUER
Baudirektor
Kreditanstalt für Wiederaufbau
Hauptabteilung Technik
Palmengarlenstr. 5-9
D-6000 Frankfurt am Main

<u>Ireland</u>

Mr. J. SHEEDY
Head, Road Construction Section
An Foras Forbartha
Pottery Road
Deansgrange
Co. Dublin

<u>Italy</u>

Ing. E. SCOTTO
Chief Engineer
ANAS
Via Monzambano 10
I-00185 Rome

<u>Japan</u>

Mr. Tomonobu NAKAOKA
Deputy Director
International Affairs Division
Ministry of Construction
2-1-3 Kasumigaseki, Chiyoda-ku
Tokyo 305

<u>Netherlands</u>

Ing. H. VAN DEN BELD
Senior Engineer
DHV Consulting Engineers
Laan 1914 Nr. 35/P.O. Box 85
3800 AB Amersfoort

Mr. A. BRESSER M.Sc.
DHV Consulting Engineers
Project Director
P.O. Box 85
3800 AB Amersfoort

<u>Norway</u>

Mr. Geir REFSDAL
Senior Engineer
Norwegian Directorate of Roads
Road Research Laboratory
P.O. Box 6390 Etterstad
N-0604 Oslo 6

<u>Turkey</u>	Mr. Mehmet A.ÖNAL Pavement Design Eng. Karayollari Genel Müdürlügü Üstyapi Subesi Müdürlügü Ankara
	Mr. A. ISFENDIYAR Ind. Eng. Karayollari Genel Müdürlügü Planlama Subesi Müdürlügü Ankara
<u>United Kingdom</u>	Mr. David BROOKS Head of Pavement Management Section of Overseas Unit TRRL Old Wokingham Road Crowthorne, Berks RG11 6AU
<u>United States</u>	Mr. John CUTRELL Director International Highway Programs (HPI-10) Federal Highway Administration Department of Transportation 400 7th Street, S.W. Washington, D.C. 20590
	Mr. B. LORD Pavement Division (HNR-20) Office of Engineering and Highway Operations Research and Development Federal Highway Administration Department of Transportation 6300 Georgetown Pike McLean, Virginia 22101
<u>World Bank</u>	Dr. W. PATERSON The World Bank 1818 H Street N.W. Washington, D.C. 20433
<u>OECD</u>	Mr. B. HORN Road Transport Research 2, rue André-Pascal 75775 Paris Cedex 16

Distributors of World Bank Publications

ARGENTINA
Carlos Hirsch, SRL
Galeria Guemes
Florida 165, 4th Floor-Ofc. 453/465
1333 Buenos Aires

AUSTRALIA, PAPUA NEW GUINEA, FIJI, SOLOMON ISLANDS, VANUATU, AND WESTERN SAMOA
D.A. Books & Journals
11-13 Station Street
Mitcham 3132
Victoria

AUSTRIA
Gerold and Co.
Graben 31
A-1011 Wien

BAHRAIN
Bahrain Research and Consultancy Associates Ltd.
P.O. Box 22103
Manama Town 317

BANGLADESH
Micro Industries Development Assistance Society (MIDAS)
House 56, Road 7A
Dhanmondi R/Area
Dhaka 1209

Branch office:
156, Nur Ahmed Sarak
Chittagong 4000

BELGIUM
Publications des Nations Unies
Av. du Roi 202
1060 Brussels

BRAZIL
Publicacoes Tecnicas Internacionais Ltda.
Rua Peixoto Gomide, 209
01409 Sao Paulo, SP

CANADA
Le Diffuseur
C.P. 85, 1501B rue Ampère
Boucherville, Quebec
J4B 5E6

CHINA
China Financial & Economic Publishing House
8, Da Fo Si Dong Jie
Beijing

COLOMBIA
Enlace Ltda.
Apartado Aereo 34270
Bogota D.E.

COSTA RICA
Libreria Trejos
Calle 11-13
Av. Fernandez Guell
San Jose

COTE D'IVOIRE
Centre d'Edition et de Diffusion Africaines (CEDA)
04 B.P. 541
Abidjan 04 Plateau

CYPRUS
MEMRB Information Services
P.O. Box 2098
Nicosia

DENMARK
SamfundsLitteratur
Rosenoerns Allé 11
DK-1970 Frederiksberg C

DOMINICAN REPUBLIC
Editora Taller, C. por A.
Restauracion e Isabel la Catolica 309
Apartado Postal 2190
Santo Domingo

EL SALVADOR
Fusades
Avenida Manuel Enrique Araujo #3530
Edificio SISA, 1er. Piso
San Salvador

EGYPT, ARAB REPUBLIC OF
Al Ahram
Al Galaa Street
Cairo

The Middle East Observer
8 Chawarbi Street
Cairo

FINLAND
Akateeminen Kirjakauppa
P.O. Box 128
SF-00101
Helsinki 10

FRANCE
World Bank Publications
66, avenue d'Iéna
75116 Paris

GERMANY, FEDERAL REPUBLIC OF
UNO-Verlag
Poppelsdorfer Allee 55
D-5300 Bonn 1

GREECE
KEME
24, Ippodamou Street Platia Plastiras
Athens-11635

GUATEMALA
Librerias Piedra Santa
Centro Cultural Piedra Santa
11 calle 6-50 zona 1
Guatemala City

HONG KONG, MACAO
Asia 2000 Ltd.
6 Fl., 146 Prince Edward Road, W.
Kowloon
Hong Kong

HUNGARY
Kultura
P.O. Box 139
1389 Budapest 62

INDIA
Allied Publishers Private Ltd.
751 Mount Road
Madras - 600 002

Branch offices:
15 J.N. Heredia Marg
Ballard Estate
Bombay - 400 038

13/14 Asaf Ali Road
New Delhi - 110 002

17 Chittaranjan Avenue
Calcutta - 700 072

Jayadeva Hostel Building
5th Main Road Gandhinagar
Bangalore - 560 009

3-5-1129 Kachiguda Cross Road
Hyderabad - 500 027

Prarthana Flats, 2nd Floor
Near Thakore Baug, Navrangpura
Ahmedabad - 380 009

Patiala House
16-A Ashok Marg
Lucknow - 226 001

INDONESIA
Pt. Indira Limited
Jl. Sam Ratulangi 37
Jakarta Pusat
P.O. Box 181

IRELAND
TDC Publishers
12 North Frederick Street
Dublin 1

ITALY
Licosa Commissionaria Sansoni SPA
Via Benedetto Fortini, 120/10
Casella Postale 552
50125 Florence

JAPAN
Eastern Book Service
37-3, Hongo 3-Chome, Bunkyo-ku 113
Tokyo

KENYA
Africa Book Service (E.A.) Ltd.
P.O. Box 45245
Nairobi

KOREA, REPUBLIC OF
Pan Korea Book Corporation
P.O. Box 101, Kwangwhamun
Seoul

KUWAIT
MEMRB Information Services
P.O. Box 5465

MALAYSIA
University of Malaya Cooperative Bookshop, Limited
P.O. Box 1127, Jalan Pantai Baru
Kuala Lumpur

MEXICO
INFOTEC
Apartado Postal 22-860
14060 Tlalpan, Mexico D.F.

MOROCCO
Societe d'Etudes Marketing Marocaine
12 rue Mozart, Bd. d'Anfa
Casablanca

NETHERLANDS
InOr-Publikaties b.v.
P.O. Box 14
7240 BA Lochem

NEW ZEALAND
Hills Library and Information Service
Private Bag
New Market
Auckland

NIGERIA
University Press Limited
Three Crowns Building Jericho
Private Mail Bag 5095
Ibadan

NORWAY
Narvesen Information Center
Bertrand Narvesens vei 2
P.O. Box 6125 Etterstad
N-0602 Oslo 6

OMAN
MEMRB Information Services
P.O. Box 1613, Seeb Airport
Muscat

PAKISTAN
Mirza Book Agency
65, Shahrah-e-Quaid-e-Azam
P.O. Box No. 729
Lahore 3

PERU
Editorial Desarrollo SA
Apartado 3824
Lima

PHILIPPINES
National Book Store
701 Rizal Avenue
P.O. Box 1934
Metro Manila

POLAND
ORPAN
Patac Kultury i Nauki
00-901 Warszawa

PORTUGAL
Livraria Portugal
Rua Do Carmo 70-74
1200 Lisbon

SAUDI ARABIA, QATAR
Jarir Book Store
P.O. Box 3196
Riyadh 11471

MEMRB Information Services
Branch offices:
Al Alsa Street
Al Dahna Center
First Floor
P.O. Box 7188
Riyadh

Haji Abdullah Alireza Building
King Khaled Street
P.O. Box 3969
Damman

33, Mohammed Hassan Awad Street
P.O. Box 5978
Jeddah

SINGAPORE, TAIWAN, MYANMAR, BRUNEI
Information Publications Private, Ltd.
02-06 1st Fl., Pei-Fu Industrial Bldg.
24 New Industrial Road
Singapore 1953

SOUTH AFRICA, BOTSWANA
For single titles:
Oxford University Press Southern Africa
P.O. Box 1141
Cape Town 8000

For subscription orders:
International Subscription Service
P.O. Box 41095
Craighall
Johannesburg 2024

SPAIN
Mundi-Prensa Libros, S.A.
Castello 37
28001 Madrid

Libreria Internacional AEDOS
Consell de Cent, 391
08009 Barcelona

SRI LANKA AND THE MALDIVES
Lake House Bookshop
P.O. Box 244
100, Sir Chittampalam A. Gardiner Mawatha
Colombo 2

SWEDEN
For single titles
Fritzes Fackboksforetaget
Regeringsgatan 12, Box 16356
S-103 27 Stockholm

For subscription orders
Wennergren-Williams AB
Box 30004
S-104 25 Stockholm

SWITZERLAND
For single titles
Librairie Payot
6, rue Grenus
Case postal 381
CH 1211 Geneva 11

For subscription orders:
Librairie Payot
Service des Abonnements
Case postal 3312
CH 1002 Lausanne

TANZANIA
Oxford University Press
P.O. Box 5299
Dar es Salaam

THAILAND
Central Department Store
306 Silom Road
Bangkok

TRINIDAD & TOBAGO, ANTIGUA BARBUDA, BARBADOS, DOMINICA, GRENADA, GUYANA, JAMAICA, MONTSERRAT, ST. KITTS & NEVIS, ST. LUCIA, ST. VINCENT & GRENADINES
Systematics Studies Unit
#9 Watts Street
Curepe
Trinidad, West Indies

TURKEY
Haset Kitapevi, A.S.
Istiklal Caddesi No. 469
Beyoglu
Istanbul

UGANDA
Uganda Bookshop
P.O. Box 7145
Kampala

UNITED ARAB EMIRATES
MEMRB Gulf Co.
P.O. Box 6097
Sharjah

UNITED KINGDOM
Microinfo Ltd.
P.O. Box 3
Alton, Hampshire GU34 2PG
England

URUGUAY
Instituto Nacional del Libro
San Jose 1116
Montevideo

VENEZUELA
Libreria del Este
Aptdo. 60.337
Caracas 1060-A

YUGOSLAVIA
Jugoslovenska Knjiga
YU-11000 Belgrade Trg Republike

ZIMBABWE
Longman Zimbabwe
P.O. Box ST 125, Southerton
Harare

WHERE TO OBTAIN OECD PUBLICATIONS
OÙ OBTENIR LES PUBLICATIONS DE L'OCDE

Argentina - Argentine
Carlos Hirsch S.R.L.
Galeria Güemes, Florida 165, 4° Piso
1333 Buenos Aires
Tel. 30.7122, 331.1787 y 331.2391
Telegram: Hirsch-Baires
Telex: 21112 UAPE-AR. Ref. s/2901
Telefax: (1)331-1787

Australia - Australie
D.A. Book (Aust.) Pty. Ltd.
11-13 Station Street (P.O. Box 163)
Mitcham, Vic. 3132 Tel. (03)873.4411
Telex: AA37911 DA BOOK
Telefax: (03)873.5679

Austria - Autriche
OECD Publications and Information Centre
4 Simrockstrasse
5300 Bonn (Germany) Tel. (0228)21.60.45
Telex: 8 86300 Bonn
Telefax: (0228)26.11.04

Gerold & Co.
Graben 31
Wien I Tel. (0222)533.50.14

Belgium - Belgique
Jean De Lannoy
Avenue du Roi 202
B-1060 Bruxelles
Tel. (02)538.51.69/538.08.41
Telex: 63220 Telefax: (02)538.08.41

Canada
Renouf Publishing Company Ltd.
1294 Algoma Road
Ottawa, Ont. K1B 3W8 Tel. (613)741.4333
Telex: 053-4783 Telefax: (613)741.5439
Stores:
61 Sparks Street
Ottawa, Ont. K1P 5R1 Tel. (613)238.8985
211 Yonge Street
Toronto, Ont. M5B 1M4 Tel. (416)363.3171

Federal Publications
165 University Avenue
Toronto, ON M5H 3B9 Tel. (416)581.1552
Telefax: (416)581.1743
Les Publications Fédérales
1185 rue de l'Université
Montréal, PQ H3B 1R7 Tel.(514)954-1633

Les Éditions La Liberté Inc.
3020 Chemin Sainte-Foy
Sainte-Foy, P.Q. G1X 3V6
Tel. (418)658.3763
Telefax: (418)658.3763

Denmark - Danemark
Munksgaard Export and Subscription Service
35, Norre Sogade, P.O. Box 2148
DK-1016 Kobenhavn K
Tel. (45 33)12.85.70
Telex: 19431 MUNKS DK
Telefax: (45 33)12.93.87

Finland - Finlande
Akateeminen Kirjakauppa
Keskuskatu 1, P.O. Box 128
00100 Helsinki Tel. (358 0)12141
Telex: 125080 Telefax: (358 0)121.4441

France
OECD/OCDE
Mail Orders/Commandes par correspondance:
2 rue André-Pascal
75775 Paris Cedex 16 Tel. (1)45.24.82.00
Bookshop/Librairie:
33, rue Octave-Feuillet
75016 Paris Tel. (1)45.24.81.67
(1)45.24.81.81
Telex: 620 160 OCDE
Telefax: (33-1)45.24.85.00

Librairie de l'Université
12a, rue Nazareth
13602 Aix-en-Provence Tel. 42.26.18.08

Germany - Allemagne
OECD Publications and Information Centre
4 Simrockstrasse
5300 Bonn Tel. (0228)21.60.45
Telex: 8 86300 Bonn
Telefax: (0228)26.11.04

Greece - Grèce
Librairie Kauffmann
28 rue du Stade
105 64 Athens Tel. 322.21.60
Telex: 218187 LIKA Gr

Hong Kong
Government Information Services
Publications (Sales) Office
Information Service Department
No. 1 Battery Path
Central Tel. (5)23.31.91
Telex: 802.61190

Iceland - Islande
Mal Mog Menning
Laugavegi 18, Postholf 392
121 Reykjavik Tel. 15199/24240

India - Inde
Oxford Book and Stationery Co.
Scindia House
New Delhi 110001 Tel. 331.5896/5308
Telex: 31 61990 AM IN
Telefax: (11)332.5993
17 Park Street
Calcutta 700016 Tel. 240832

Indonesia - Indonésie
Pdii-Lipi
P.O. Box 269/JKSMG/88
Jakarta12790 Tel. 583467
Telex: 62 875

Ireland - Irlande
TDC Publishers - Library Suppliers
12 North Frederick Street
Dublin 1 Tel. 744835/749677
Telex: 33530 TDCP EI Telefax : 748416

Italy - Italie
Libreria Commissionaria Sansoni
Via Benedetto Fortini, 120/10
Casella Post. 552
50125 Firenze Tel. (055)645415
Telex: 570466 Telefax: (39.55)641257
Via Bartolini 29
20155 Milano Tel. 365083
La diffusione delle pubblicazioni OCSE viene assicurata dalle principali librerie ed anche da:
Editrice e Libreria Herder
Piazza Montecitorio 120
00186 Roma Tel. 679.4628
Telex: NATEL I 621427

Libreria Hoepli
Via Hoepli 5
20121 Milano Tel. 865446
Telex: 31.33.95 Telefax: (39.2)805.2886

Libreria Scientifica
Dott. Lucio de Biasio "Aeiou"
Via Meravigli 16
20123 Milano Tel. 807679
Telefax: 800175

Japan- Japon
OECD Publications and Information Centre
Landic Akasaka Building
2-3-4 Akasaka, Minato-ku
Tokyo 107 Tel. 586.2016
Telefax: (81.3)584.7929

Korea - Corée
Kyobo Book Centre Co. Ltd.
P.O. Box 1658, Kwang Hwa Moon
Seoul Tel. (REP)730.78.91
Telefax: 735.0030

Malaysia/Singapore - Malaisie/Singapour
University of Malaya Co-operative Bookshop Ltd.
P.O. Box 1127, Jalan Pantai Baru 59100
Kuala Lumpur
Malaysia Tel. 756.5000/756.5425
Telefax: 757.3661

Information Publications Pte. Ltd.
Pei-Fu Industrial Building
24 New Industrial Road No. 02-06
Singapore 1953 Tel. 283.1786/283.1798
Telefax: 284.8875

Netherlands - Pays-Bas
SDU Uitgeverij
Christoffel Plantijnstraat 2
Postbus 20014
2500 EA's-Gravenhage Tel. (070)78.99.11
Voor bestellingen: Tel. (070)78.98.80
Telex: 32486 stdru Telefax: (070)47.63.51

New Zealand -Nouvelle-Zélande
Government Printing Office
Customer Services
P.O. Box 12-411
Freepost 10-050
Thorndon, Wellington
Tel. 0800 733-406 Telefax: 04 499-1733

Norway - Norvège
Narvesen Info Center - NIC
Bertrand Narvesens vei 2
P.O. Box 6125 Etterstad
0602 Oslo 6
Tel. (02)67.83.10/(02)68.40.20
Telex: 79668 NIC N Telefax: (47 2)68.53.47

Pakistan
Mirza Book Agency
65 Shahrah Quaid-E-Azam
Lahore 3 Tel. 66839
Telex: 44886 UBL PK. Attn: MIRZA BK

Portugal
Livraria Portugal
Rua do Carmo 70-74
1117 Lisboa Codex Tel. 347.49.82/3/4/5

Singapore/Malaysia Singapour/Malaisie
See "Malaysia/Singapore"
Voir "Malaisie/Singapour"

Spain - Espagne
Mundi-Prensa Libros S.A.
Castello 37, Apartado 1223
Madrid 28001 Tel. (91) 431.33.99
Telex: 49370 MPLI Telefax: (91) 275.39.98

Libreria Internacional AEDOS
Consejo de Ciento 391
08009 -Barcelona Tel. (93) 301-86-15
Telefax: (93) 317-01-41

Sweden - Suède
Fritzes Fackboksföretaget
Box 16356, S 103 27 STH
Regeringsgatan 12
DS Stockholm Tel. (08)23.89.00
Telex: 12387 Telefax: (08)20.50.21
Subscription Agency/Abonnements:
Wennergren-Williams AB
Box 30004
104 25 Stockholm Tel. (08)54.12.00
Telex: 19937 Telefax: (08)50.82.86

Switzerland - Suisse
OECD Publications and Information Centre
4 Simrockstrasse
5300 Bonn (Germany) Tel. (0228)21.60.45
Telex: 8 86300 Bonn
Telefax: (0228)26.11.04

Librairie Payot
6 rue Grenus
1211 Genève 11 Tel. (022)731.89.50
Telex: 28356

Maditec S.A.
Ch. des Palettes 4
1020 Renens/Lausanne Tel. (021)635.08.65
Telefax: (021)635.07.80

United Nations Bookshop/Librairie des Nations-Unies
Palais des Nations
1211 Genève 10
Tel. (022)734.60.11 (ext. 48.72)
Telex: 289696 (Attn: Sales)
Telefax: (022)733.98.79

Taïwan - Formose
Good Faith Worldwide Int'l. Co. Ltd.
9th Floor, No. 118, Sec. 2
Chung Hsiao E. Road
Taipei Tel. 391.7396/391.7397
Telefax: (02) 394.9176

Thailand - Thalande
Suksit Siam Co. Ltd.
1715 Rama IV Road, Samyan
Bangkok 5 Tel. 251.1630

Turkey - Turquie
Kültur Yayinlari Is-Türk Ltd. Sti.
Atatürk Bulvari No. 191/Kat. 21
Kavaklidere/Ankara Tel. 25.07.60
Dolmabahce Cad. No. 29
Besiktas/Istanbul Tel. 160.71.88
Telex: 43482B

United Kingdom - Royaume-Uni
H.M. Stationery Office
Gen. enquiries Tel. (01) 873 0011
Postal orders only:
P.O. Box 276, London SW8 5DT
Personal Callers HMSO Bookshop
49 High Holborn, London WC1V 6HB
Telex: 297138 Telefax: 873.8463
Branches at: Belfast, Birmingham, Bristol, Edinburgh, Manchester

United States - États-Unis
OECD Publications and Information Centre
2001 L Street N.W., Suite 700
Washington, D.C. 20036-4095
Tel. (202)785.6323
Telex: 440245 WASHINGTON D.C.
Telefax: (202)785.0350

Venezuela
Libreria del Este
Avda F. Miranda 52, Aptdo. 60337
Edificio Galipan
Caracas 106
Tel. 951.1705/951.2307/951.1297
Telegram: Libreste Caracas

Yugoslavia - Yougoslavie
Jugoslovenska Knjiga
Knez Mihajlova 2, P.O. Box 36
Beograd Tel. 621.992
Telex: 12466 jk bgd

Orders and inquiries from countries where Distributors have not yet been appointed should be sent to: OECD Publications Service, 2 rue André-Pascal, 75775 Paris Cedex 16.
Les commandes provenant de pays où l'OCDE n'a pas encore désigné de distributeur devraient être adressées à : OCDE, Service des Publications, 2, rue André-Pascal, 75775 Paris Cedex 16.

Printed on behalf of the World Bank
and the Organisation for Economic Co-operation and Development by
OECD PUBLICATIONS 2, rue André-Pascal 75775 PARIS CEDEX 16
PRINTED IN FRANCE
(77 89 05 1) ISBN 92-64-13309-7 No. 44923 1990
ISBN 0-8213-1391-6

ROAD TRANSPORT RESEARCH

ROAD MONITORING FOR MAINTENANCE MANAGEMENT

VOLUME 1

manual for developing countries

REPORT PREPARED BY AN OECD SCIENTIFIC EXPERT GROUP IN CO-OPERATION WITH THE WORLD BANK

The findings, interpretations, and recommendations in this study are the results of joint research by the OECD's Road Transport Research Programme and the World Bank. The report has been approved by the 23 governments participating in the Steering Committee of the OECD Road Transport Research Programme, i.e. all OECD Member governments except New Zealand. The report, however, does not necessarily reflect the views of the World Bank, its Board of Executive Directors, nor the countries they represent, other than those represented on the OECD RTR Steering Committee.

WORLD BANK
ORGANISATION FOR ECONOMIC CO-OPERATION AND DEVELOPMENT

Library of Congress Cataloging-in-Publication Data :

ROAD MONITORING MANUAL FOR MAINTENANCE MANAGEMENT.

Contents: Vol. 1. Manual for developing countries - Vol. 2. Damage catalogue for developing countries. 1. Roads - Developing countries - Maintenance and repair - Management. 2. Pavements - Developing countries - Maintenance and repair - Management.

I. ORGANISATION FOR ECONOMIC CO-OPERATION AND DEVELOPMENT. II. INTERNATIONAL BANK FOR RECONSTRUCTION AND DEVELOPMENT.

TE220.R594 1990 625.7'6'091724 89-70979
ISBN 92-64-13309-7 (set)
ISBN 0-8213-1391-6 (World Bank : set)

•

Pursuant to article 1 of the Convention signed in Paris on 14th December 1960, and which came into force on 30th September 1961, the Organisation for Economic Co-operation and Development (OECD) shall promote policies designed:

- to achieve the highest sustainable economic growth and employment and a rising standard of living in Member countries, while maintaining financial stability, and thus to contribute to the development of the world economy;
- to contribute to sound economic expansion in Member as well as non-member countries in the process of economic development; and
- to contribute to the expansion of world trade on a multilateral, non-discriminatory basis in accordance with international obligations.

The original Member countries of the OECD are Austria, Belgium, Canada, Denmark, France, the Federal Republic of Germany, Greece, Iceland, Ireland, Italy, Luxembourg, the Netherlands, Norway, Portugal, Spain, Sweden, Switzerland, Turkey, the United Kingdom and the United States. The following countries became Members subsequently through accession at the dates indicated hereafter: Japan (28th April 1964), Finland (28th January 1969), Australia (7th June 1971) and New Zealand (29th May 1973).

The Socialist Federal Republic of Yugoslavia takes part in some of the work of the OECD (agreement of 28th October 1961).

Publié en français sous le titre:

SUIVI DES ROUTES POUR LA GESTION DE L'ENTRETIEN

VOLUME 1

MANUEL POUR
LES PAYS EN DÉVELOPPEMENT

FOREWORD

The OECD Road Transport Research Programme has two main fields of activity:

-- The international co-operation in road and road transport research to provide scientific support for decisions by governments and international governmental organisations, and to assess future strategies concerning roads and road transport problems and the priority policy concerns of OECD Member countries;

-- The information and documentation programme (IRRD -- International Road Research Documentation), a co-operative scheme that provides a mechanism for the systematic world-wide exchange of information on scientific literature and current research programmes.

In order to assist development aid agencies and highway administrations in third world countries, one of the priorities of the Programme's scientific and technical activities is on the maintenance management of road infrastructure and the evaluation of traffic safety measures and strategies in developing countries.

This report in the form of a manual -- together with a separate publication containing the Damage Catalogue -- should help in improving maintenance practice and cost effectiveness of road maintenance operations in developing countries.

DEDICATION

This Manual is dedicated to Per Fossberg of the World Bank and David Brooks of the Overseas Unit of the U.K. Transport and Road Research Laboratory, both of whom died in 1989, in recognition of their outstanding contribution to international co-operation and to advancing road technology and maintenance in the developing world.

ABSTRACT

IRRD No. 818900

The OECD Scientific Expert Group on "Pavement Monitoring Systems for Developing Countries" was created to prepare, in co-operation with the World Bank, a manual for road pavement inspection applicable to Third World countries' conditions. The report starts with the Background and Needs of this activity and, apart from a list of definitions, has six chapters -- Introduction; Establishing a Road Monitoring Scheme; Road Condition Survey; Detailed Visual Inspection; Data Processing and Storage; Training -- and four annexes -- Basic Instruction for Implementing Inspection Activities, Assessment of Road Surface Roughness (Evenness), Summary of Damage Catalogue, Assessment Using Special Measuring Equipment. The full Damage Catalogue is issued in a separate publication.

Subject classification: 10, 60

Fields: Economics and administration; Maintenance

Keywords: Surveillance, manual (book), highway, developing countries, administration, data acquisition, data processing, education, personnel, surfacing, evenness, damage, inventory, apparatus (measuring), evaluation (assessment), deterioration, road network

ACKNOWLEDGEMENTS

This manual was prepared by an OECD Road Transport Research Group with major input from the following experts: Messrs. A. Bresser and H. van den Beld (Netherlands), D. Brooks (United Kingdom), L. Francken (Belgium), M. Gorski (Belgium), H. Neubauer (Germany), W. Paterson (World Bank), C. Peyronne (France) and G. Refsdal (Chairman, Norway) and with the support of the OECD Secretariat and the World Bank. For the material received we should also like to thank the following institutions: Cadic Gombert, Consulting Engineers -- Belgium; Centre de Recherches Routières -- Belgium; DHV Consulting Engineers -- Netherlands; Forschungsgesellschaft für Strassen- und Verkehrswesen -- Germany; Laboratoire Central des Ponts et Chaussées -- France; Ministère des Transports et des Travaux Publics -- Mali; M & R International, Consulting Engineers -- Belgium; Overseas Unit, TRRL -- United Kingdom; Société d'ingénierie BCEOM -- France.

BACKGROUND AND NEEDS

ROAD TRANSPORT IN DEVELOPING COUNTRIES

Road transport in developing countries is an important sector of economic activity and investment, and functions as a catalyst to the overall economic and social development process. The role of road transport is relatively more important in most developing countries than in OECD Member countries because of limitations in other sectors of their national transport systems. In particular, rural road accessibility is one of the key elements of small-holder development; most developing countries have agriculturally-based economies and road transport plays therefore an essential role in marketing agricultural (village) production. Road infrastructure is likewise important for providing access to health, education and agricultural extension services as well as the distribution of agricultural inputs such as fertilizer and pesticides.

Fast growth of national vehicle fleets and a road construction boom during the 1960's and 1970's have been common characteristics of past developments in road transport in developing countries. The significance of road transport in the developing world becomes clear when different transport modes are compared. In a majority of cases, where either railways or waterway transport have long provided the backbone of the national transport system, the market shares of these traditional modes have declined, while road-based services have dramatically increased. In those developing countries which never had a railway system and where there is no natural infrastructure for inland navigation, the growth of the road transport sector is even more important for the countries' evolution from subsistence economies to market economies. In fact, it is not at all exceptional in developing countries today, for road transport to provide 80 to 90 per cent or more of the total inland and/or border crossing traffic. Infrastructure maintenance and improvement are therefore vital to overall economic and social progress.

The fast expanding road haulage industry, combined with limited private car ownership, has led to a very high proportion of heavy, pavement-damaging freight vehicle traffic. In addition there is a pronounced tendency to overload vehicles which increases the damage to pavements and road structures. Whereas in OECD countries construction of heavier vehicles was accompanied by improvements in the bearing capacity of infrastructure, there is a growing gap between the use of increasingly heavier freight vehicles and inadequate pavement and bridge structures in developing countries.

During the boom years of road transport development in developing countries, priority was given to new construction and the remaining funds were insufficient for proper infrastructure maintenance. Furthermore, the small share of available resources which was used for maintenance purposes was often applied ineffectively. Over the years the limited resources devoted to the upkeep of road networks together with the growth of heavy freight traffic have created a large backlog of road rehabilitation needs. The result is escalating road repair and vehicle operating costs as well as safety problems. Also, the capital invested in the transport fleet is used uneconomically. This is regrettable, because the budgetary constraints in developing countries demand the best possible use of available resources.

ROAD CONDITIONS

The state of road infrastructure is revealed by available World Bank statistics. In the 85 countries to which the World Bank has provided assistance for road development, 26 per cent of paved roads (some 270 000 km) and 32 per cent of unpaved main roads outside of urban areas are in a "poor" condition and have already deteriorated to the point of needing reconstruction. Only around 30 per cent of the main road network is in fair condition (1).

The World Bank figures are striking also from another point of view, as shown in Table 1. There is a relatively high percentage of road infrastructure in "fair condition" (an estimated 42 per cent of paved roads and 37 per cent of unpaved main roads of the network under consideration), which is indicative of urgent maintenance needs "if these roads are to be saved from declining to poor condition". The term "fair condition" (by definition) suggests the need for resurfacing within the near future (1).

By and large, the actual status of major road networks (not including unclassified rural roads) is alarming throughout the different development regions, as shown in Table 1. Though more difficult to quantify, experience in developing countries suggests that insufficient importance is given to rural road networks. Numerous countries in Africa, Asia and Latin America have vast networks of mostly low volume earth and gravel roads, which often account for more than 80 per cent of the entire road system.

MAJOR SHORTCOMINGS AND CONCERNS

The wide lack of systematic and appropriate road maintenance in developing countries is mainly a result of a lack of funds and a developed maintenance organisation. Only recently has the importance of systematic monitoring, network evaluation and regular maintenance works become generally recognised. This trend is shown in a reduction in the scale of new road

---

1. World Bank. Road Deterioration in Developing Countries: Causes and Remedies. Policy Paper. Washington, D.C., 1988.

Table 1

CONDITION OF MAIN ROAD NETWORKS BY DEVELOPMENT REGION

	Paved Roads			Unpaved Roads		
Region	Good	Fair	Poor	Good	Fair	Poor
Eastern Africa	42	32	26	42	30	28
Western Africa	52	23	25	20	36	44
East Asia	20	59	21	41	34	25
South Asia	18	45	36	6	39	55
Europe, Middle East and N. Africa	40	35	25	30	46	24
Latin America and Caribbean	44	32	24	24	43	33
Average	32	42	26	31	37	32

Source: World Bank, (1988).

construction and, in turn, an increased allocation of funds for maintenance projects and works in many developing countries. Particular attention is now given to the use of local materials and efforts directed towards increasing the bearing capacity of the road network by appropriate techniques.

However, the funding available after budgetary reallocation from new road investment to infrastructure maintenance remains often insufficient.

Larger budgets and supportive economic and financial policies certainly help ease the situation. However, it would be misleading to assume that desirable road maintenance improvements can be achieved simply by allocating sufficient funds to rehabilitation and maintenance works. In many developing countries, a lack of professional manpower, know-how and experience at both the central level of decision-making and the operational level, constitute a serious constraint to appropriate repair and maintenance of the road network. Attempts are being made to meet the need for education and training and to improve the efficiency of road maintenance organisations. In fact, international and bilateral assistance programmes increasingly emphasise infrastructure maintenance technical assistance.

As one important step in that direction, a three-part manual on road maintenance, produced as a result of collaboration between government organisations in France, Germany, and Great Britain, was introduced at the Third African Highway Maintenance Conference held in Addis Ababa in 1982. The guide books, which were produced for the United Nations Economic Commission for Africa, aim primarily at the operational level (road foremen), and present simple maintenance practices for remedying defects in paved and unpaved roads and structures as developed through many years of experience obtained on road projects.

These handbooks will be complemented by this manual which addresses the maintenance engineer as the key person in the maintenance organisation at the district (and operational) level and on the decision maker at the central level of developing countries.

Accordingly, this manual provides the outline of a system so that maintenance needs can be identified simply and easily. Though many developing countries have been working hard to upgrade facilities and to attain appropriate technologies and administrative procedures best suited to local circumstances, there are still ample shortcomings to be overcome through either sustained national initiatives, international co-operation and/or bilateral agency support, in order to effectively carry out maintenance activities:

-- At the right time;
-- Of the right quality;
-- At the right place.

ECONOMIC AND SOCIAL IMPACT OF A PROPERLY MAINTAINED ROAD SYSTEM

Proper road maintenance contributes to reliable transport at reduced cost as there is a direct link between pavement condition and vehicle operating costs (VOC). The International Road Federation has stated that "because of poor road maintenance the road sector in many developing countries is operating well below its optimum and this has a direct effect on many sectors of the national economy". VOC savings accrue mainly to the state and private haulage industries and, via reductions in transport fees, to consumers.

It should be noted that the share of the distribution cost in the total cost, defined as the difference between factory and sales price, (including handling, storage and associated costs) of products in developing countries may be 50 per cent or more. Generally, the major portion of such intermediate costs are attributable to freight expenses. It is therefore vital that the transport cost component be minimised.

Road improvements and accessibility stimulate increased agricultural production and crop diversification due to the improvement of marketing opportunities. They promote rural industrialisation and socio-economic integration and stimulate the economy. The importance, of course, of these multiple effects varies with the adopted improvement standard and also depends on the degree of complementary development activities.

In addition to these aspects associated with a properly maintained road system, there is also the question of protecting a large capital investment (the highway system) in a manner that will maximise the benefits of the investment relative to the costs of protection, or in this case, maintenance. It is clear that timely and appropriate maintenance can slow the rate of deterioration, delay the need for major, expensive, repairs such as reconstruction, and generally extend the service life of a road to attain the highest economic benefit.

Appropriate maintenance investments can produce major economic benefits as well as reducing user costs. As the above-cited World Bank report suggests, $41 billion (1984 prices) is needed to meet the reconstruction needs for paved and unpaved main roads outside of urban areas in some 85 countries to which the World Bank has provided road development assistance. This contrasts sharply with the $12 billion cost of upkeep programmes which, if undertaken at the appropriate time, could have saved the roads. Again this is independent of huge vehicle operating cost savings. The cost of meeting the current and future maintenance needs of these countries, if done in a timely and economical manner, will be one-fifth to one-third of the cost of addressing the problems if they are handled as they were in the past.

Finally, it should be noted that an improperly maintained road system can present increased safety hazards to the highway user. These can lead to more accidents with their associated costs of human death and injuries, property damage, etc., as well as human suffering. A proper road monitoring system will identify existing and potential safety hazards which an adequate maintenance programme will eliminate.

This manual is a step in the process of improving the preservation of road networks. It is not a maintenance management system nor a pavement management system. However, it does constitute an important component, indeed the first step of gathering the information for such systems. This manual presents simple, usable guidelines for data acquisition on paved and unpaved roads. As such it is hoped that the manual will be used to make timely and economic investments to preserve the integrity and serviceability of road networks.

TABLE OF CONTENTS

Page

DEFINITIONS 14

Chapter I

INTRODUCTION 19

I.1. Purpose 19
I.2. Scope of maintenance 20
I.3. Road monitoring 20
I.4. Road maintenance organisation 21
I.5. Structure and content of the manual 22

Chapter II

ESTABLISHING A ROAD MONITORING SCHEME 25

II.1. Framework 25
II.2. Office preparations 25
II.2.1. General 25
II.2.2. Functional classification of the road network .. 26
II.2.3. Division into sections and sub-sections 27
II.2.4. Referencing of sections and sub-sections 29
II.3. Organisation 31
II.3.1. Basic principles 31
II.3.2. Preparation of worksheets 32
II.3.3. Composition and organisation of inspection teams 35
II.3.4. Documents 35
II.3.5. Equipment 36

Chapter III

ROAD CONDITION SURVEY 39

III.1. Purpose 39
III.2. Method for evaluation 39
III.3. Survey procedures 39
III.3.1. Forms 39
III.3.2. Parameters 41
III.4. Implementation 44
III.4.1. Frequency 44
III.4.2. Preparations before the survey 45
III.4.3. The two phases of the survey 45

Page

III.5. Assessment of results 46
III.5.1. Summary of survey results 46
III.5.2. Interpretation of survey results 46

Chapter IV

DETAILED VISUAL INSPECTION 51

IV.1. Purpose 51
IV.2. Method of evaluation 51
IV.3. Inspection procedures 51
IV.3.1. Forms 51
IV.3.2. Parameters for unpaved roads 52
IV.3.3. Parameters for paved roads 55
IV.4. Implementation 55
IV.4.1. Frequency 55
IV.4.2. Preparations before the inspection 55
IV.4.3. Method 56
IV.5. Assessment of results 56
IV.5.1. Summary of inspection results 56
IV.5.2. Interpretation of inspection results 56

Chapter V

DATA AGGREGATION AND ANALYSIS 59

V.1. Data processing and storage 59
V.1.1. Validating the data 59
V.1.2. Filing the data 60
V.1.3. Main types of data processing 61
V.2. Use of data for evaluating pavement condition and work programmes 65
V.2.1. Overall monitoring of pavement condition 65
V.2.2. Specifying the jobs needed 67

Chapter VI

TRAINING 69

VI.1. General 69
VI.1.1. Identification of training needs 69
VI.1.2. Training needs analysis 70
VI.1.3. Planning of the training programme 70
VI.1.4. Execution of the training programme 70
VI.1.5. Evaluation of the results 71
VI.2. Job description 71
VI.2.1. Job description of the district engineer 71
VI.2.2. Job description of the road inspector 71
VI.3. Outline for training courses 72
VI.3.1. Training of the DE 73
VI.3.2. Training of the RI 73

Page

VI.4. Trainee selection 74
VI.5. Training location 74
VI.6. Conclusions 74
Appendix 75

Annex A

BASIC INSTRUCTIONS FOR IMPLEMENTING INSPECTION ACTIVITIES 79

A.1. Itinerary 79
A.2. Road condition survey (RCS) 79
A.2.1. Parameters to be evaluated 89
A.2.2. Survey procedure 90
A.2.3. Summary of results 90
A.3. Detailed visual inspection (DVI) 90
A.3.1. Parameters for "unpaved roads", Form VI 90
A.3.2. Parameters for "paved roads", Form VII 92
A.3.3. Inspection procedure 92
A.3.4. Summary of results 92

Annex B

ASSESSMENT OF ROAD SURFACE ROUGHNESS (EVENNESS) 95

B.1. International roughness index: IRI 95
B.2. Subjective determination of the IRI 95

Annex C

SUMMARY OF DAMAGE CATALOGUE 101

C.1. Introduction 101
C.2. Damage sheets 101
C.3. Rating of damage 103
C.3.1. Road condition survey (RCS) 103
C.3.2. Detailed visual inspection (DVI) 105

Annex D

OVERVIEW OF SPECIAL MEASURING EQUIPMENT 107

D.1. Rationalising and enhancing visual inspections 107
D.2. High output equipment for measuring roughness 108
D.3. High output equipment for measuring structural performance 109
D.3.1. Structural integrity: deflections 109
D.3.2. Individual layers 109
D.4. Equipment for measuring skid resistance 110
D.5. Final remark 110

LIST OF GROUP MEMBERS 111

DEFINITIONS

The list hereunder presents a selection of the most important notions. Its purpose is to aid users of the manual and to clarify some terms which might be open to interpretation. In no case should this list be considered exhaustive.

AADT

-- Annual average daily traffic (veh./day)

BASE

-- See pavement courses.

BEARING CAPACITY (load structural capacity)

-- Characterises the structural condition of a pavement related to the pavements' capacity to withstand future traffic loads;

-- Expressed by indicators that reflect the response of the pavement to loads -- deflection -- or by indicators of pavement design and construction (e.g. thickness of pavement courses);

-- Can be measured by different devices (See Annex D).

CARRIAGEWAY

-- See cross-section.

CONDITION, pavement

-- Is concerned with the functional as well as structural conditions of the pavement.

CONDITION RATING

-- Assigns a value to the functional condition of a road (pavement) often partly based on its structural condition.

COURSES, pavement

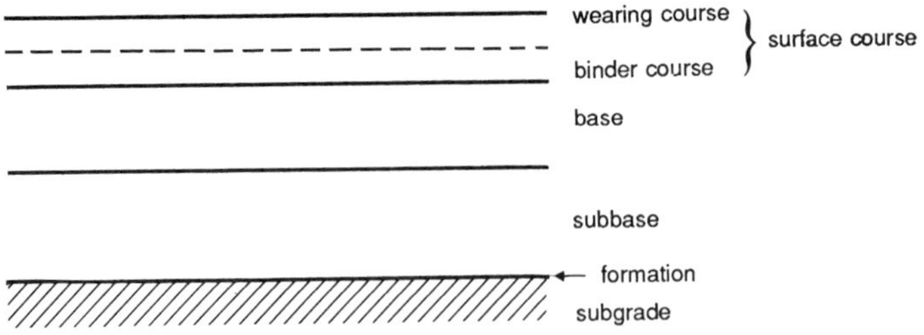

CROSS-SECTION: Vertical section through the road at right angles to the centre line; the main components are:

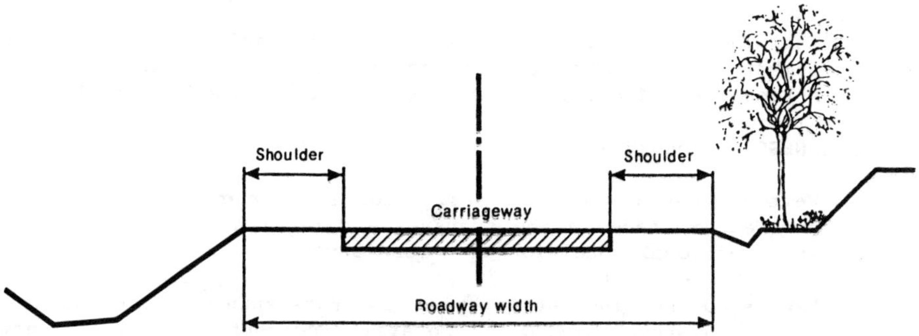

DAMAGE (defect, deficiency, distress, deterioration)

-- See Annex C and Damage Catalogue for precise definitions of the various types illustrated with photographs of typical examples, suitable inspection methods, rating criteria, most probable causes and factors and possible remedial maintenance activities.

DEFLECTION

-- The transient vertical displacement of a pavement surface under applied loading.

DETERIORATION, pavement

-- The wear and damage processes under which the functional and structural capability of the pavement decreases;

-- Leads to defects, distress or damage.

EQUIVALENT AXLE LOAD

-- Expressed in terms of a standard single-axle load of 80 or 100 kN.

EVALUATION (appraisal) of road (pavement) condition

-- The process of quantifying road (pavement) condition by means of an indicator;

-- Involves measurements and/or observations, comparisons with thresholds and incorporation in rating systems.

EVENNESS, longitudinal (longitudinal roughness)

-- Vertical deviation from the designed shape, occurring at the pavement surface in the longitudinal direction of the road affecting the functional condition of a road (pavement);

-- Expressed by indicators such as slope variance, roughness, etc. or incorporated into rating systems giving overall concepts of the (functional) condition;

-- Measured by a variety of devices, e.g. equipment that measures the number and magnitude of vertical deviations between the body of an automobile and the centre of the rear axle (See Annexes B and D).

EVENNESS, transverse

-- Vertical deviations from designated shape e.g. ruts occurring at the pavement surface in the transverse direction of the road, affecting the functional condition of a road pavement;

-- Expressed by indicators such as rut-depth or incorporated into rating systems giving an overall concept of the (functional) condition;

-- Measured with a straight edge, by transverse profile measuring equipment or by road surface testing devices obtaining cross-profiles (see Annex D and previous OECD reports).

FORMATION

-- See pavement courses.

FUNCTIONAL CONDITION (performance condition)

-- Determines how well the pavement performs its function as a riding surface for vehicular traffic (riding quality, driving comfort);

-- Relates to longitudinal evenness, transverse evenness, skid resistance, surface texture, etc.;

-- Expressed by means of indicators or by overall rating systems designed to measure driving comfort, serviceability, etc.

INDICATORS (of condition)

-- Designates parameters used to quantify road (pavement) condition;

-- Usually defined through the measurement and/or inspection methods employed.

INTERVENTION, level of

-- A defined value (threshold) of a condition indicator, which when exceeded requires that maintenance be undertaken;

-- Compare level of warning.

MAINTENANCE

-- Any activity undertaken to preserve the original surface and structural qualities of a road or a pavement;

-- Deferred maintenance, functional maintenance, periodic maintenance, preventive maintenance, programmed maintenance, remedial maintenance, routine maintenance, structural maintenance, surface maintenance;

-- See OECD reports on this subject.

MAINTENANCE MANAGEMENT SYSTEM

-- The process of organising, scheduling and controlling maintenance activities in order to make the best possible use of resources available.

MONITORING

-- Methods used to collect data on the structural and functional condition of a road or pavement.

NETWORK LEVEL

-- Refers to decisions applicable to the entire highway network under the jurisdiction of a road authority.

PAVEMENT MANAGEMENT SYSTEM (PMS)

-- There is at present no universally acceptable definition. In this report, and in contrast to a maintenance management system, the PMS is defined as the process of co-ordinating, planning and programming a comprehensive set of activities -- including construction, maintenance and rehabilitation -- in order to minimise the whole life cost of the road.

RATING SYSTEM

-- A procedure to rate numerically pavement (road) condition on the basis of measurements and/or observations, weighting factors, etc.

ROUGHNESS, longitudinal (longitudinal evenness)

-- See longitudinal evenness.

SERVICEABILITY

-- The level of quality of a road (pavement) directly perceptible to road users.

SKID RESISTANCE

-- Attainable friction between the road surface and the tyres of vehicles affecting the functional condition of a road (pavement);

-- Expressed by indicators such as braking force coefficient, sideways force coefficient, skid number (present normalised gradient), skid number (speed gradient), etc.;

-- Measured by a variety of devices (See Annex D and previous OECD reports).

THRESHOLD (trigger value)

-- See intervention and/or warning level.

WARNING, level of

-- A defined value (threshold) of a pavement condition indicator that when exceeded advises that maintenance be considered;

-- Compare with level of intervention;

WHEEL TRACKS

-- The area of a traffic lane in which the wheels of vehicles usually travel.

REFERENCES

1. OECD. ROAD RESEARCH. Maintenance techniques for road surfacings. Organisation for Economic Co-operation and Development. Paris, 1978.

2. OECD. ROAD RESEARCH. Catalogue of road surface deficiencies. Organisation for Economic Co-operation and Development. Paris, 1978.

3. PIARC. Technical dictionary of road terms. Permanent International Association of Road Congresses. 5th Ed. Paris, 1982.

4. OECD. ROAD RESEARCH. Road surface characteristics. Organisation for Economic Co-operation and Development. Paris, 1984.

5. OECD. ROAD TRANSPORT RESEARCH. Pavement management systems. Organisation for Economic Co-operation and Development. Paris, 1987.

6. OECD/CRR. Maintenance of unpaved roads in developing countries. Centre de recherches routières (CRR). Brussels, 1987.

Chapter I

INTRODUCTION

I.1. PURPOSE

Regular and reliable information on road condition is essential for managing the maintenance of a road network, both for assessing the physical and financial needs and for evaluating the effectiveness of road maintenance practices. The road monitoring method described here is intended to aid the management of road maintenance and can also be used for rehabilitation. It is a basic, operational method; it is neither complex, nor dependent on sophisticated mechanical equipment and highly-trained individuals to carry out the work.

The method is applicable to engineered, paved and unpaved roads, including the drainage facilities which are an important priority for maintenance. Engineering structures, bridges (*), and tunnels, are not included here because of the unique nature of the technical information required for them.

A road monitoring method needs to be consistent with local requirements and resources. For this reason the method proposed here does not present a rigid scheme, but may be adjusted to suit local conditions and preferences. It does comprise the components that should be common to all monitoring methods, and thus is intended to serve as a basic guide. It should be stressed however, that the system presented, while potentially a highly effective tool, should not be used blindly; it must be supported by the knowledge and experience of the maintenance engineer.

(*) The OECD Road Transport Research Programme has produced a number of relevant publications on bridges, including "Bridge Inspection", "Bridge Maintenance" and "Bridge Rehabilitation and Strengthening". Presently, a study on "Bridge Management" is being conducted. Also, the U.K. Transport and Road Research Laboratory's Overseas Unit has recently issued "A Guide to Bridge Inspection and Data Systems for District Engineers" and a "Bridge Inspector's Handbook".

I.2. SCOPE OF MAINTENANCE

At the start it may be useful to define certain basic terms. Road maintenance as defined in the U.N. Road Maintenance Handbook consists of "routine and periodic activities to keep pavements, shoulders, slopes, drainage and all other structures and property within the right of way as nearly as possible in their as-constructed or renewed condition. Maintenance includes minor repairs and improvements to eliminate the cause of the defect and avoid unnecessary repetition of maintenance efforts".

Certain maintenance activities are performed routinely (normally at intervals of one year or less) and these are defined as routine maintenance. Other operations are performed less frequently (intervals of more than one year) and are periodic maintenance. A third type of maintenance activity, urgent or emergency maintenance, should be carried out immediately upon identification, and is not included here.

I.3. ROAD MONITORING

Management of a road network requires different information at different levels of decision-making, for example, for planning, for programming, for design, and for implementation. The data to be collected by an inspection system, and where, when and how it should be collected, depend very largely on the use of the data. The two principal management applications are:

-- Maintenance management *, whose aim is the efficient organisation, scheduling, and budget-control of maintenance activities within the budget or fiscal year; in formal systems this often includes assessment of the effectiveness of the works and the productivity rates;

-- Pavement management -- a planning and programming procedure -- that minimizes the whole life cost of the road and requires information on the condition and trafficking of the pavement, in order to evaluate and schedule appropriate major maintenance or rehabilitation works, in both medium-term plans and yearly programmes.

In both management processes, there are two indispensible components, information and a decision process. The information component comprises a monitoring or data collection method and a data bank for storing the

---

* In view of the complexity of the institutional issues and rapid progress in management techniques, the OECD has decided to launch a joint international project with the aim of developing an up-to-date road maintenance management guide; in the framework of this study it is planned to evaluate the present Monitoring Manual with pilot-tests in several developing regions.

information, and provides quantitative data on the state of the road network. The information elements may be either simple and basic, or detailed and comprehensive, depending on the resources and capabilities of the road administration. The decision processes may range from simple standardized criteria and repair prescriptions, to advanced analytical techniques, but these may also evolve over time and it is wise to ensure that the information subsystem is capable of being expanded or upgraded without fundamental change.

The road monitoring method presented here is intended to serve as a basic information subsystem, and is presented in two levels:

- The primary level provides basic information for maintenance management through a Road Condition Survey (RCS), which would be conducted over the entire network under the road authority's jurisdiction at an approximately annual frequency. The basic survey is primarily visual and may be supplemented by some instrumented measurements. It provides a regular assessment of road condition and needs, and could be used as a basis on which to orientate budget allocations and to evaluate the effectiveness of maintenance practices.
- The second level is a Detailed Visual Inspection (DVI) of road condition, which might be done selectively for only those sections appearing to need major works. It is intended to identify the sections to be considered for the programming of major maintenance, but, typically, such works require additional information, particularly a structural evaluation of the pavement and measure of traffic loading, and these require instrumented measurements which are not treated in detail here. Thus, if a formal pavement management system exists or is implemented in the future, the DVI should be supplemented by other types of investigation if the rehabilitation is to be properly defined.

It is important to formulate an overall plan for the approach to road management in which the context of the monitoring system -- i.e. the RCS and the DVI -- can be identified. Ideally, this should commence at a reasonably basic level and be upgraded as resources permit. A key consideration is that the monitoring system be sustainable by, and appropriate to, the resources available and the institutional capacity of the road administration.

I.4. ROAD MAINTENANCE ORGANISATION

The objective of a road maintenance organisation is to create the environment, that is, the conditions and logistical support, for the effective implementation of all road maintenance activities. The organisation should be simple, straight-forward and, most importantly, meet the specific needs, requirements and resources of a country. Normally the organisation consists of three organisational levels: the central level which establishes national highway plans and budgets; the regional level where maintenance requirements are defined, planning, execution and supervision of maintenance is carried out, and, budget control and cost administration occurs; and the district level where the maintenance activities, including monitoring, are carried out under the supervision of maintenance engineers.

While this is a standard organisational structure for a maintenance organisation, the monitoring system presented in this manual does not assume a particular organisational structure. The preparatory work discussed in Chapter II, for example, normally will be done at the district level, however there are no problems if it is done partially at the regional or central and partially at the district levels.

I.5. STRUCTURE AND CONTENT OF THE MANUAL

A general scheme of the monitoring system is presented in Figure I.1. This figure will be used repeatedly throughout the manual as elements of the system are introduced so that a logical progression through the process can be identified. The numbers on the left hand side of the flow chart boxes correspond to the section of the manual where a particular element of the monitoring system is described.

Chapter II discusses the preparatory work entailed in establishing a road monitoring scheme. Based on the assumption that a location reference system and a road inventory have been set up, the Chapter reviews office preparations, organisation of the monitoring scheme including preparation of work sheets and other documents needed as well as equipment requirements. Depending on the basic organisational set up this work may be accomplished either at a central or a district level.

Chapters III and IV are the heart of the manual since they describe the monitoring process: the way it is undertaken, the parameters evaluated and the forms used to record the data. The procedure entails two levels: i) the Road Condition Survey (RCS) (Chapter III) in which the general condition of the road network is assessed, sections are identified for further, more detailed inspection and damage requiring immediate attention is detected; and ii) the Detailed Visual Inspection (DVI) (Chapter IV) where those sections identified in the Road Condition Survey requiring additional attention are evaluated to determine the type, extent and severity of damage, thus enabling the planning of maintenance operations.

In conjunction with these Chapters, the manual contains annexes which are relevant to the execution of the monitoring. Annex A provides the basic instructions for implementing inspection activities illustrated by an example; Annex B describes the International Roughness Index (IRI), and how IRI can be used with other data to give a global evaluation of the carriageway quality for use as a decision criterion in the monitoring system; Annex C summarises the Damage Catalogue issued in a separate booklet and which describes with illustrations and text the most frequently occurring deficiencies so that these may be identified and rated; for the RCS, damage for road parameters and structures is categorised, and for DVI, a distinction is made between damage of unpaved and paved roads and additional information is provided on the probable causes of the damage and proposed maintenance activities.

General considerations regarding application of the data collected are provided in Chapter V including data processing and storage. The use of data for evaluating pavement condition and work programmes is also illustrated.

Figure I.1. **FLOW CHART OF THE ROAD MONITORING SYSTEM AND ITS INTERACTION WITH PLANNING AND PROGRAMMING MAINTENANCE ACTIVITIES**

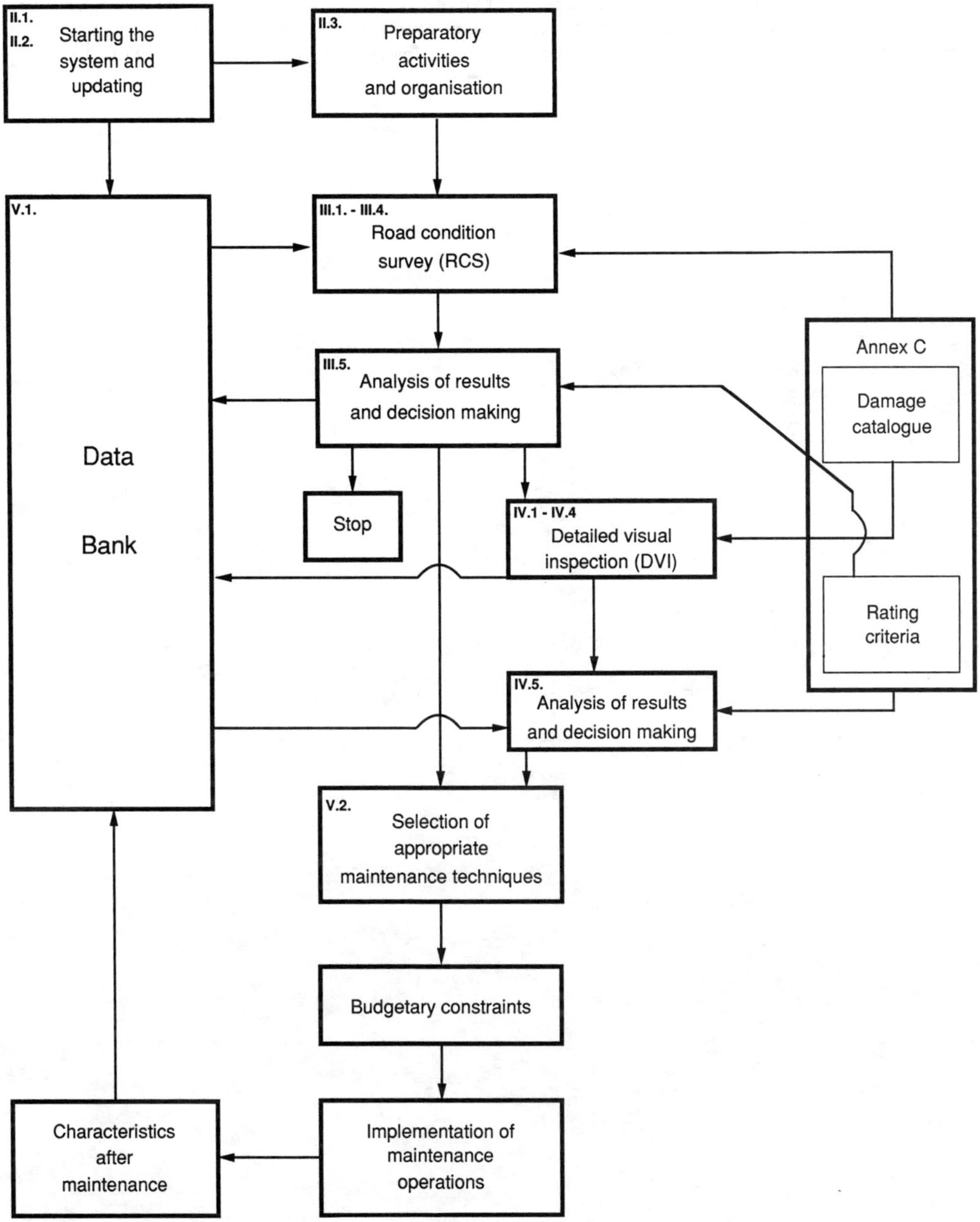

Although the monitoring programme does not require extensive training, a training programme is of critical importance so as to obtain consistent and reliable data, as outlined in Chapter VI.

Advanced measuring devices are also not required. However high output measuring instruments can enhance efficiency and reduce subjectivity and inaccuracy of data. Annex D provides an overview of some selected high output equipment which has proven effective and economic in a number of OECD highway agencies.

Chapter II

ESTABLISHING A ROAD MONITORING SCHEME

II.1. FRAMEWORK

The road monitoring scheme developed in this manual assumes that a location reference system and a road inventory are in place.

The basic requirements for a well-organised data collection system are the following:

-- A division of the road network into sections and sub-sections. These should be as homogeneous as possible.

-- Every section and sub-section must be identified by a simple code comprising road, section and sub-section number.

-- A standardized system of data collection.

The road monitoring scheme requires a special introduction within the already existing road management system. Project groups and survey teams have to be organised and instructed before the monitoring process can be started. The introduction must aim at familiarising all persons involved with the importance and the potential benefits of the system.

A timetable for the monitoring process has to be determined. All parties involved in the operation of the system must schedule their contribution in such a way that the information required for decision making will become available to the management at the appropriate time.

II.2. OFFICE PREPARATIONS

II.2.1. General

The road monitoring system as described in this manual is of a general nature. It allows for adjustments to the specific demands of a particular maintenance organisation or subdivision of a road network.

The basic organisational set-up (see Section I.4.), will determine to a large extent the level(s) at which the monitoring will be planned, managed, staffed and carried out. This may depend on the type of the maintenance management adopted (centralised, decentralised or mixed) in addition to many other factors including cost considerations, budget planning needs and other key management controls. In this manual it is assumed that it is used in a maintenance district, however, the procedure will be the same if used for a larger road network.

The following elements are dealt with in this Chapter:

-- Functional characteristics of the road network;

-- Division of the road network into sections and sub-sections;

-- The referencing of sections and sub-sections.

II.2.2. Functional classification of the road network

A prerequisite to dividing a road network into sections and sub-sections is a knowledge of the functional characteristics of the different parts of the network. Depending on these characteristics the maintenance requirements will be different.

The following criteria will be used to classify the function of a road:

-- Traffic load, i.e. annual average daily traffic (AADT) or, if available, the daily number of heavy vehicles.

-- Road category, i.e. primary, secondary, tertiary roads, or similar.

The introduction of a functional indicator reflecting the type of pavement or other factors may be appropriate if such a system already exists in a country. However, in the absence of such criteria, Table II.1 provides an example of suitable functional criteria.

Table II.1

AN EXAMPLE OF A FUNCTIONAL EVALUATION OF ROADS

Criteria	Class	Functional indicator	
		Paved roads	Unpaved roads
Traffic (AADT)*	1	> 1 000	> 200
	2	200 - 1 000	50 - 200
	3	< 200	< 50
Road category	A	Primary or main road	
	B	Secondary or feeder road	
	C	Tertiary or local road	

* Annual average daily traffic.

The traffic figures should reflect high, medium and low traffic volumes. In many countries the figures presented in Table II.1 do not reflect realistic traffic conditions. In such cases, more appropriate classes of traffic volumes or road categories should be introduced.

II.2.3. Division into sections and sub-sections

Before the road network can be divided into sections, the road classification has to be indicated on a large scale map, i.e. a topographical map with a scale of 1 : 100 000. See Figure II.1 for an example of the preparation of such a map.

The first large scale map must be transformed into a simplified road map (see Figure II.2). This simplified road map is used to divide the roads into sections and sub-sections (see Figure A.4). The following information should be included on the maintenance district's simplified road map:

- -- Functional indicator of the roads;
- -- District boundaries or change of authority (i.e. highway authority versus district authority);
- -- Important engineering structures;
- -- Location of towns and villages;
- -- Location of sections and sub-sections.

Figure II.1. **EXAMPLE OF THE DIVISION OF A MAINTENANCE DISTRICT ROAD NETWORK INTO SUB-SECTIONS**

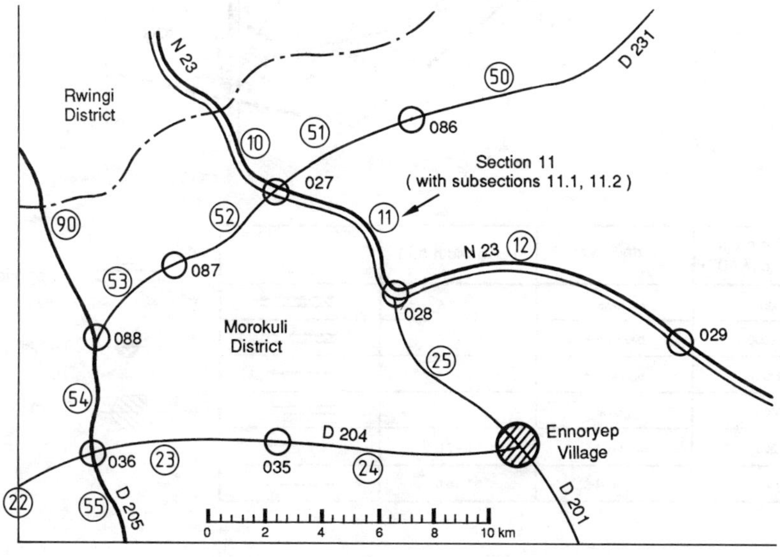

Figure II.2. **SIMPLIFIED ROAD MAP OF MOROKULI MAINTENANCE DISTRICT**

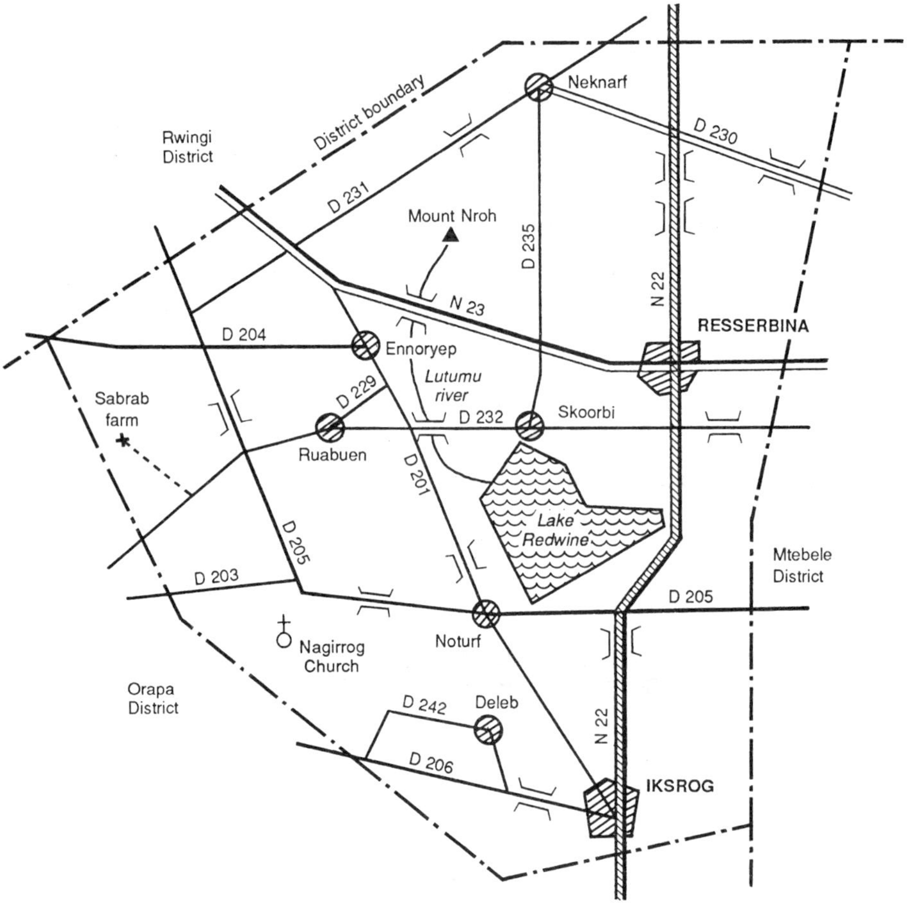

KEY

Annual average daily traffic (AADT)	Surface type	Road width (m.)	
> 1000	bituminous	7.0 -7.5	
200 - 1000	bituminous	5.5 - 7.0	
< 200	bituminous	5.0 - 5.5	
> 200	gravel	7.0	
50 - 200	gravel	5.0 - 7.0	
< 50	gravel	3.5 - 5.0	

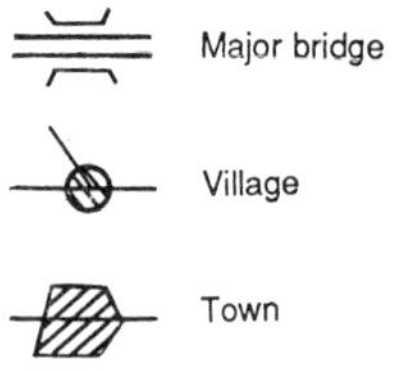

Subsections should be homogeneous in character with regard to:

-- Cross-section;
-- Pavement structure (as far as is recognisable);
-- Environment (e.g. cut, fill, open country, forest, village, gradient, etc.).

The length of the sub-sections should not be less than 500 m unless there are special reasons for the introduction of shorter sub-sections. On the other hand, the length of the sub-sections should not exceed 5 km. Subsections should preferably conform with those sections that are in use for other purposes, such as traffic counts, accident surveys, etc.

The length of sections depends on the length of the sub-sections. A section should consist of no more than five sub-sections, but here also exceptions to the general rule are allowed. Sections should be chosen in a logical way, i.e. between villages, major intersections, bridges, etc.

Inspection will be facilitated if strip maps of the sections and sub-sections (see Figures II.3 and A.5) are prepared. These maps can give a lot of information about sections and sub-sections and ensure better inspection of culverts and other small structures. Strip maps should contain among other information the following:

-- Location of culverts and bridges;
-- Location of quarries and borrow pits and the length of access roads to quarries and borrow pits;
-- Location of intersections;
-- Chainage.

It should be emphasised that the road monitoring system described is very flexible and can accomodate wide variations in the selection of sections and sub-sections.

II.2.4. Referencing of sections and sub-sections

Limits of sections and sub-sections have to be referenced in the field. The most widely used referencing system is a system that uses position marks and distances from position marks. This system allows for the easy location of the data collected in the field and updating of the system can easily be done without affecting areas not involved in the updating.

The position marks may be special items like existing marker posts, road intersections, a bridge, culvert, etc.

If these road elements are too far apart or if a marker is required at a location where no appropriate road element is available, special marker posts should be placed. This may be the case where the road environment changes or there is a significant gradient change.

Position markers must be easily recognisable in the field and sufficiently robust to deter vandals and remain permanent.

Figure II.3. **STRIP MAP OF ROAD N23**

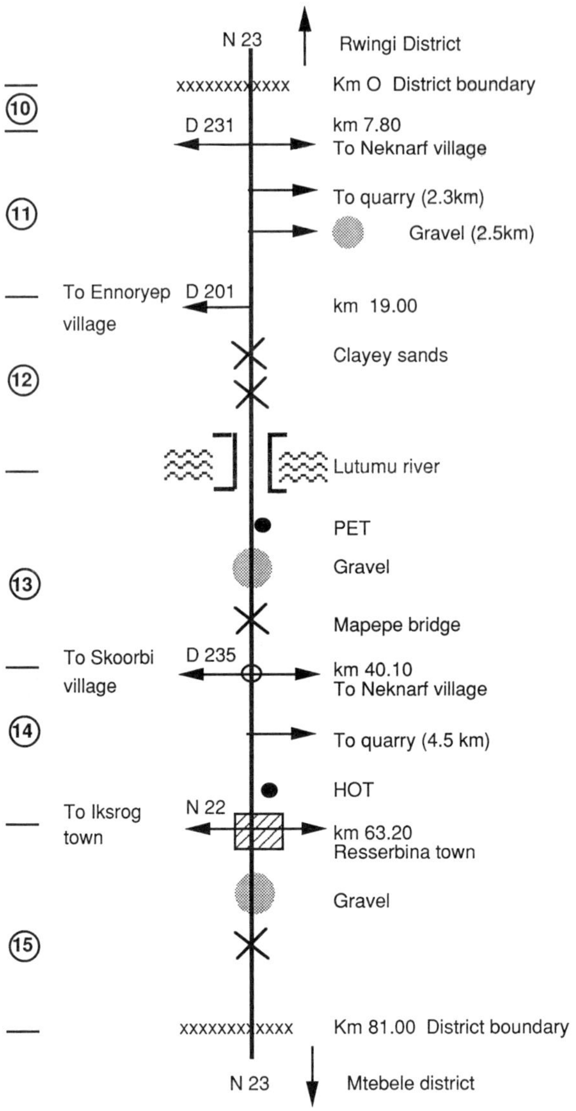

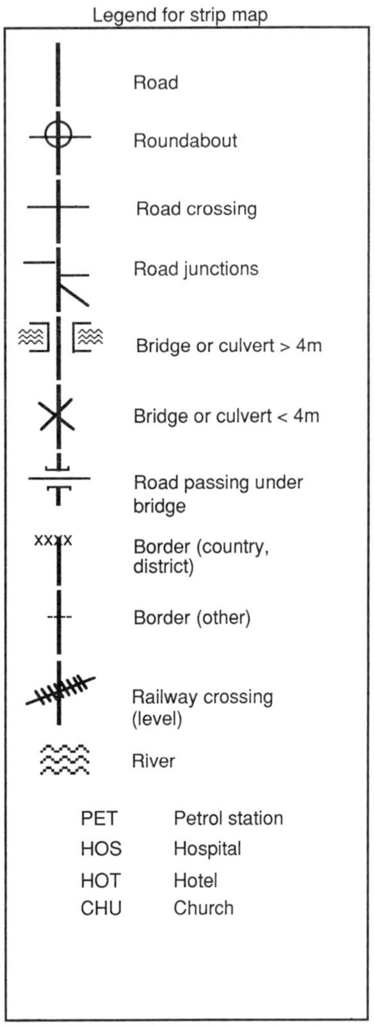

II.3. ORGANISATION

II.3.1. Basic principles

The monitoring scheme is based on two different levels of inspection:

a) The "road condition survey" (RCS);

b) The "detailed visual inspection" (DVI).

The conditions appropriate to the implementation of both types of survey are described in the following paragraphs and further developed in Chapters III and IV. The inspection procedures presented are suitable for paved and unpaved roads.

Usually, the RCS is to be conducted over the entire road network, preferably at least annually, because it provides a regular assessment of all maintenance needs and of the condition of the road network. The DVI must be conducted on at least those road sections that were identified by the RCS as needing major carriageway maintenance.

If there are no constraints on time and resources, the two types of survey could be used over the entire network under consideration. If both surveys are not usually done on the entire network, then the quality of decision-making can be improved in the long-term by monitoring a set of representative sections for research. The sections are sampled throughout the network and selected on the basis of age, traffic, structure and environment in order to provide long term performance evaluation data which could be useful for modelling road deterioration.

Figure II.4 illustrates how the preparatory work for organising the road condition survey and detailed visual inspection fits into the overall scheme.

Figure II.4. **POSITION OF PREPARATORY ACTIVITIES IN THE GENERAL SCHEME**

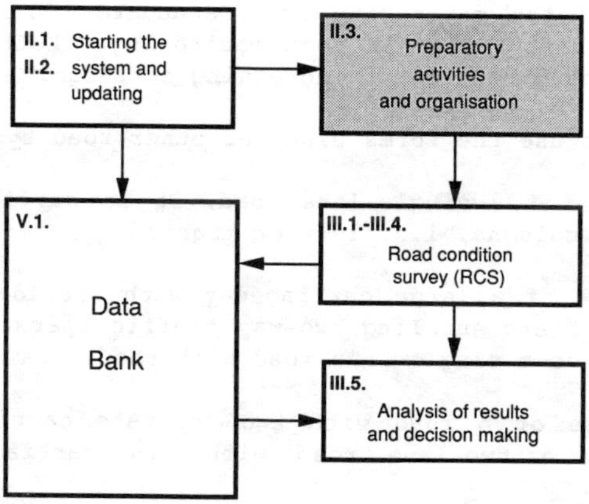

In order to perform inspections in the most economical way, daily itineraries should be prepared by the district office. Sections have to be combined logically. Travel time to and from the office or lodging site and travel time between sections should be taken into consideration when planning the daily itinerary. In general, a daily itinerary for a road condition survey may be of a total length of 30-50 km. For detailed visual inspections the daily production may amount to approximately 5-8 km depending on the complexity of the sub-sections to be inspected and on the travel time involved. It is recommended that actual output to be expected be determined by the engineer.

II.3.2. Preparation of worksheets

The itinerary of each road condition survey and of each detailed visual inspection may be given by means of a completed Form I. See Figure II.5 for an example of this form.

Also, the upper blocks of the road condition survey forms, Form II, are completed in the district office. See Figure II.6 for an example of this form.

The documentation required for the inspection is completed with a map of the area and, if available, strip maps of the roads to be inspected.

As identification points (IP on Form I) the following abbreviations may be used:

- L = Limit of maintenance district
- J = Junction
- C = Crossing
- B = Bridge
- T = Town
- V = Village
- CC = Change of cross-section
- CT = Change of terrain type
- CV = Culvert
- SM = Special marker

The forms for the road condition survey and the detailed visual inspection are based on the monitoring of a standard two-lane road in both directions (Right = R, Left = L). In road monitoring, "Left" and "Right" are defined by driving in the direction of increasing chainage.

It is possible to use the forms also for other road types:

i) In the case of a single-lane road, it is sufficient to complete one of the columns, i.e. left or right;

ii) In the case of a large carriageway with shoulders, surfaced in a single width and enabling two-way traffic operation, the road can be treated as a single-lane road with shoulders [see i) above];

iii) In the case of a road with two separate carriageways, they are treated as a two-lane road with its carriageways considered separately.

Figure II.5. **FORM FOR SURVEY (RCS) OR INSPECTION (DVI) ITINERARY ("FORM I")**

MAINTENANCE DISTRICT: Completed by:	Itinerary No.:	PAGE ◯

ROUTE		MAIN SECTIONS					
Identification points (I.P.)		Road		No. of main section	Chai-nage (km)	Length	
Kind	Designation	Class	No.			Single (km)	Added (km)
1	2	3	4	5	6	7	8
				◯			
				◯			
				◯			
				◯			
				◯			
				◯			
				◯			
				◯			
				◯			
				◯			

Sub-sections											No. of main section
Number of sub-sections	◯ From	To	◯ From	To	◯ From	To	◯ From	To	◯ From	To	
IP											
Station											
Length (km)											
IP											
Station											
Length (km)											
IP											
Station											
Length (km)											
IP											
Station											
Length (km)											
IP											
Station											
Length (km)											
IP											
Station											
Length (km)											
IP											
Station											
Length (km)											
IP											
Station											
Length (km)											
IP											
Station											
Length (km)											
IP											
Station											
Length (km)											

Figure II.6. **ROAD CONDITION SURVEY (RCS) FORM ("FORM II")**

ROAD CONDITION SURVEY PAVED AND UNPAVED ROADS			ROAD DENOMINATION / CLASS:		Main Section No.
			FROM:		Km:
			TO:		Km:
OFFICE:	DISTRICT		Pavement type:	Pavement width: m	Section length: Km
Name of inspector:	:		Summary (Avg. Condition)	Carriageway:	
Date:	Weather o Clear o Rainy	Carr.way: o Dry o Drying o Wet		Road side elements:	
				Remarks:	

SUB-SECTION	◯		◯		◯		◯		◯	
Chainage of sub-section										
Roadside	L	R	L	R	L	R	L	R	L	R
Carriageway	1 2 4 3 5	1 2 4 3 5	1 2 4 3 5	1 2 4 3 5	1 2 4 3 5	1 2 4 3 5	1 2 4 3 5	1 2 4 3 5	1 2 4 3 5	1 2 4 3 5
Prevailing damage										

	Damage type		L	R	L	R	L	R	L	R	L	R
ROAD SIDE ELEMENTS	Shoulder	Deformation RCS1	1 2 3	1 2 3	1 2 3	1 2 3	1 2 3	1 2 3	1 2 3	1 2 3	1 2 3	1 2 3
		Scour RCS2	1 2 3	1 2 3	1 2 3	1 2 3	1 2 3	1 2 3	1 2 3	1 2 3	1 2 3	1 2 3
	Side drains	Siltation RCS3	1 2 3	1 2 3	1 2 3	1 2 3	1 2 3	1 2 3	1 2 3	1 2 3	1 2 3	1 2 3
		Scour RCS4	1 2 3	1 2 3	1 2 3	1 2 3	1 2 3	1 2 3	1 2 3	1 2 3	1 2 3	1 2 3
	Debris/Vegetation Encroachment RCS5		1 2 3	1 2 3	1 2 3	1 2 3	1 2 3	1 2 3	1 2 3	1 2 3	1 2 3	1 2 3
	Obstacles/ Obstructions RCS6		1 2 3	1 2 3	1 2 3	1 2 3	1 2 3	1 2 3	1 2 3	1 2 3	1 2 3	1 2 3
			1 2 3	1 2 3	1 2 3	1 2 3	1 2 3	1 2 3	1 2 3	1 2 3	1 2 3	1 2 3
			1 2 3	1 2 3	1 2 3	1 2 3	1 2 3	1 2 3	1 2 3	1 2 3	1 2 3	1 2 3
	Average Condition (Others)		+ =		+ =		+ =		+ =		+ =	
			: =		: =		: =		: =		: =	

ROADSIGNS/FURNITURE	Chainage										
	Dirty RCS7	1 2 3	1 2 3	1 2 3	1 2 3	1 2 3	1 2 3	1 2 3	1 2 3	1 2 3	1 2 3
	Damaged RCS7	1 2 3	1 2 3	1 2 3	1 2 3	1 2 3	1 2 3	1 2 3	1 2 3	1 2 3	1 2 3
	Missing RCS7	1 2 3	1 2 3	1 2 3	1 2 3	1 2 3	1 2 3	1 2 3	1 2 3	1 2 3	1 2 3
		1 2 3	1 2 3	1 2 3	1 2 3	1 2 3	1 2 3	1 2 3	1 2 3	1 2 3	1 2 3

REMARKS ____

____

____

____

II.3.3. Composition and organisation of inspection teams

The number of inspection teams will depend on the required frequency of the surveys and the way the inspection teams are organised, whether in a central or a district level organisation.

Also, the need to have the monitoring data available within a certain time for administrative reasons may influence the number of teams required. In most cases, one inspection team in each maintenance district will be sufficient.

For the initial planning of the inspection team's work, the previously mentioned production figures will be useful (road condition survey 30-50 km/day, detailed visual inspection 5-8 km/day). After some test runs the production figures may require adjustment to local conditions.

For day-to-day operations, the inspection team is composed of a driver, one or two observers and, ideally, a maintenance engineer. During inspections on roads with high traffic volumes and in densely populated areas, adequate warning signs are required and in particular during detailed visual inspections, the assistance of a signalling team may be required.

The tasks of the different members of the team are:

Driver: Responsible for driving the vehicle and the transport of the equipment. He must have some knowledge of automobile mechanics and be able to drive a four wheel drive vehicle under all conditions likely to be encountered.

Observer(s): Responsible for the inspections. The navigator of the team fills in the inspection forms and assumes all the administrative duties of the operations. He must possess an adequate technical background and be acquainted with road maintenance problems.

The observer will have followed an intensive road inspection training course (see Chapter VI). This training course deals with the type of damage to be expected, the criteria to apply, and the use of the forms, etc. Practical experience will be gained through field exercises.

Maintenance engineer: The presence of a maintenance engineer will in general only be required during a detailed visual inspection. The maintenance engineer must have a good practical knowledge of road maintenance. It is possible (depending on the extent of the maintenance district) that the observer is also a maintenance engineer.

When inspections have to be carried out in remote places or on long road sections, equipping the inspection team with camping equipment to allow for an overnight stay in the field is recommended.

II.3.4. Documents

Permanent Inspection Team Documents

-- Inspection procedure instructions (Annex A);

-- Operational procedures for the inspection methods (Annex B);

-- Damage catalogue;

-- Inspection forms.

Documents Pertaining to Each Mission

The documentation for each inspection mission is prepared by the organisation responsible. The documentation includes the following elements:

-- A detailed description of the mission (road condition survey or detailed visual inspection) and itinerary to be followed;

-- Detailed maps of the area to be inspected;

-- A list of road sections and sub-sections to be inspected: this will be recorded on the itinerary form ("Form I"), the format of which is presented in Figure II.5;

-- The itinerary can also be described in a schematic form as shown in the strip map, see Figure II.3;

-- The RCS and/or DVI inspection forms for the different road sections to be inspected. One form will be used for each section and each type of survey/inspection. The general information and the inventory data will, in advance, have been completed on each form by the office responsible.

II.3.5. Equipment

The following equipment is required for the field inspection:

-- A vehicle large enough to transport all equipment and to protect the team during bad weather. For unpaved roads within the network a four wheel drive vehicle is highly recommended.

-- Safety warning devices: boards, cones, rotating hazard lights for the vehicle and flags; fluorescent waistcoats for staff.

-- A measuring wheel.

-- A measuring chain (50 m).

-- Straight edge (2 m).

-- A calibrated wedge.

-- A camber board with level.

-- Shovel.

-- A pick-axe.

-- A camera.

This is the basic required inspection equipment. A two-way radio can also be useful.

This equipment may be extended by more sophisticated instruments (see Annex D), where available, in order to acquire data that complement the DVI.

Chapter III

ROAD CONDITION SURVEY

III.1. PURPOSE

The purpose of the Road Condition Survey (RCS) is to acquire data on the general condition of the road network under the responsibility of the maintenance organisation. This survey:

a) Assesses the general condition of the road network;

b) Identifies those sections or sub-sections of the road network which are in a critical condition and those requiring maintenance or structural overlay within the near future. In the latter case, a detailed visual inspection (DVI) is mandatory (Chapter IV).

c) Detects severe damage that requires immediate maintenance action.

The relation of the Road Condition Survey to other elements in the management scheme is shown in Figure III.1.

III.2. METHOD OF EVALUATION

Quantitative figures have to be allocated to the parameters included in the RCS. This quantitative appraisal is done on the basis of the information and procedures given in the damage catalogue (Sections 1 and 2, see also Annex C.3.1, Tables C.1 and C.2). Except for the evaluation of the carriageway condition (Annex B), all road condition survey parameters are assessed on a three point scale, according to the general principles of Table III.1.

III.3. SURVEY PROCEDURES

III.3.1. Forms

Before the survey begins, its itinerary (Form I -- Figure II.5) has to be prepared and the road identification details have to be completed at the top of the Road Condition Survey Form (Form II -- Figure II.6), as described in Chapter II.

Figure III.1. **POSITION OF ROAD CONDITION SURVEY IN THE GENERAL SCHEME**

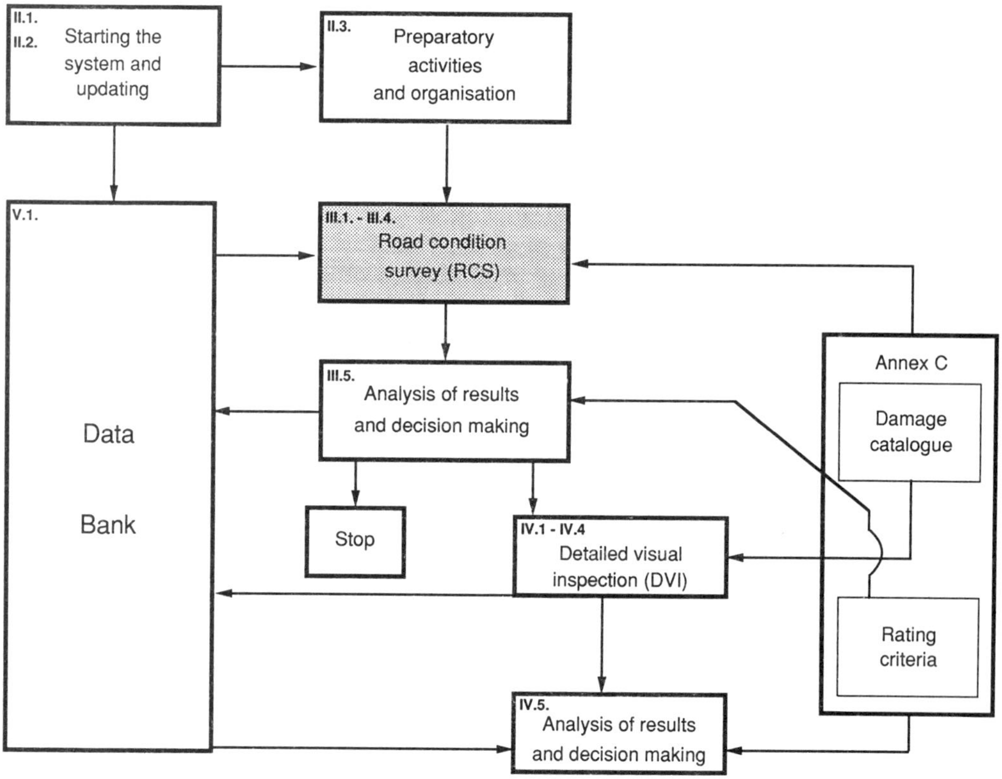

Table III.1

THREE POINT EVALUATION SCALE FOR RCS PARAMETERS

Value	Damage		General Condition
	Extent	Severity	
1	Non existent	None	Good
2	Frequent	Average	Average
3	General	Severe	Poor

III.3.2. Parameters

At this level of monitoring the parameters to be evaluated are the same for paved and unpaved roads. These parameters are listed on the Road Condition Survey Form (Form II, represented in this Chapter as Figure III.2).

Four groups are identified:

a) The carriageway (one parameter, to be evaluated on a five point scale);

b) The roadside components/elements as well as obstructions; six parameters to be evaluated on a three point scale);

c) Road signs and furniture (to be evaluated on a three point scale);

d) Road structures (culverts, bridges, small structures such as retaining walls, etc.), three parameters to be evaluated on a three point scale.

Only groups a), b) and c) are included on Form II (Figure III.2); for road structures on Form III (Figure III.3) must be used.

Evaluation of the Carriageway

This parameter gives an overall evaluation of the surface characteristic by converting the user's general impression of the riding quality of the road and visual impression of the road surface into a number.

By applying the procedure described in detail in Annex B, (Class 4), this inspection will result in a value of one to five for the surfacing of the road. These numbers are combined into an average indicating the condition of the carriageway of the section. The average prevailing damage for a sub-section is entered on the form by referring to the code given in the Damage Catalogue.

This parameter needs particular attention as it will ultimately determine the need for a detailed visual inspection.

Roadside Components/Elements

The inspection of the side elements is important in helping to keep the carriageway and the road structure itself (side drains, embankments, etc.) in an acceptable condition.

Any object or material encroaching on the road and obstructing the flow of traffic should also be recorded (vegetation, debris, etc.).

The six applicable characteristics, listed on the survey form, will be evaluated quantitatively on a three point scale: 1 = good, 2 = average and 3 = poor. This scale is described in the Damage Catalogue.

Figure III.2. **ROAD CONDITION SURVEY (RCS) FORM ("FORM II")**

ROAD CONDITION SURVEY PAVED AND UNPAVED ROADS			ROAD DENOMINATION / CLASS:		Main Section No.
			FROM:		Km:
			TO:		Km:
OFFICE:	DISTRICT		Pavement type:	Pavement width: m	Section length: Km
Name of inspector:	:		Summary (Avg. Condition)	Carriageway:	
Date:	Weather o Clear o Rainy	Carr.way: o Dry o Drying o Wet		Road side elements:	
				Remarks:	

SUB-SECTION	○	○	○	○	○
Chainage of sub-section					

Roadside	L	R	L	R	L	R	L	R	L	R
Carriageway	1 2 4 3 5	1 2 4 3 5	1 2 4 3 5	1 2 4 3 5	1 2 4 3 5	1 2 4 3 5	1 2 4 3 5	1 2 4 3 5	1 2 4 3 5	1 2 4 3 5
Prevailing damage										

		Damage type	L	R	L	R	L	R	L	R	L	R
ROAD SIDE ELEMENTS	Shoulder	Deformation RCS1	1 2 3	1 2 3	1 2 3	1 2 3	1 2 3	1 2 3	1 2 3	1 2 3	1 2 3	1 2 3
		Scour RCS2	1 2 3	1 2 3	1 2 3	1 2 3	1 2 3	1 2 3	1 2 3	1 2 3	1 2 3	1 2 3
	Side drains	Siltation RCS3	1 2 3	1 2 3	1 2 3	1 2 3	1 2 3	1 2 3	1 2 3	1 2 3	1 2 3	1 2 3
		Scour RCS4	1 2 3	1 2 3	1 2 3	1 2 3	1 2 3	1 2 3	1 2 3	1 2 3	1 2 3	1 2 3
	Debris/Vegetation Encroachment RCS5		1 2 3	1 2 3	1 2 3	1 2 3	1 2 3	1 2 3	1 2 3	1 2 3	1 2 3	1 2 3
	Obstacles/ Obstructions RCS6		1 2 3	1 2 3	1 2 3	1 2 3	1 2 3	1 2 3	1 2 3	1 2 3	1 2 3	1 2 3
			1 2 3	1 2 3	1 2 3	1 2 3	1 2 3	1 2 3	1 2 3	1 2 3	1 2 3	1 2 3
			1 2 3	1 2 3	1 2 3	1 2 3	1 2 3	1 2 3	1 2 3	1 2 3	1 2 3	1 2 3
	Average Condition (Others)		+ =		+ =		+ =		+ =		+ =	
			: =		: =		: =		: =		: =	

ROADSIGNS/FURNITURE	Chainage											
	Dirty	RCS7	1 2 3	1 2 3	1 2 3	1 2 3	1 2 3	1 2 3	1 2 3	1 2 3	1 2 3	1 2 3
	Damaged	RCS7	1 2 3	1 2 3	1 2 3	1 2 3	1 2 3	1 2 3	1 2 3	1 2 3	1 2 3	1 2 3
	Missing	RCS7	1 2 3	1 2 3	1 2 3	1 2 3	1 2 3	1 2 3	1 2 3	1 2 3	1 2 3	1 2 3
			1 2 3	1 2 3	1 2 3	1 2 3	1 2 3	1 2 3	1 2 3	1 2 3	1 2 3	1 2 3

REMARKS: ______________________________

____

____

____

Figure III.3. **CONDITION SURVEY FORM FOR STRUCTURES ("FORM III")**

ROAD CONDITION SURVEY (RCS)

OFFICE:		DISTRICT:	PAGE
Name of inspector:			
Date:	Weather o Clear o Rainy		

STRUCTURES			
Road denomination/Class:			
Type of structure	Main section	Sub-section	Chainage

CONDITION SURVEY						
Silted/ Blocked		Scour		Structural Damage		Notes/Remarks
L	R	L	R	L	R	
1 2 3	1 2 3	1 2 3	1 2 3	1 2 3	1 2 3	
1 2 3	1 2 3	1 2 3	1 2 3	1 2 3	1 2 3	
1 2 3	1 2 3	1 2 3	1 2 3	1 2 3	1 2 3	
1 2 3	1 2 3	1 2 3	1 2 3	1 2 3	1 2 3	
1 2 3	1 2 3	1 2 3	1 2 3	1 2 3	1 2 3	
1 2 3	1 2 3	1 2 3	1 2 3	1 2 3	1 2 3	
1 2 3	1 2 3	1 2 3	1 2 3	1 2 3	1 2 3	
1 2 3	1 2 3	1 2 3	1 2 3	1 2 3	1 2 3	
1 2 3	1 2 3	1 2 3	1 2 3	1 2 3	1 2 3	
1 2 3	1 2 3	1 2 3	1 2 3	1 2 3	1 2 3	
1 2 3	1 2 3	1 2 3	1 2 3	1 2 3	1 2 3	
1 2 3	1 2 3	1 2 3	1 2 3	1 2 3	1 2 3	
1 2 3	1 2 3	1 2 3	1 2 3	1 2 3	1 2 3	

All information recorded in the block called "Roadside elements" will be combined into an average condition. This average condition is obtained by computing the average arithmetic value of the values attributed to the parameters described above. This is done for both sides of the road, resulting in one average value per sub-section. These averages per sub-section are subsequently expressed in the average condition value per section, whose value is entered in the summary block.

Road Signs and Furniture; Road Markings

All items related to road safety are included as road furniture e.g. guard rails on bridges and along ravines, etc. A three point scale with the sub-division "dirty, damaged or missing" has been devised to assess the road furniture on the form.

Under "remarks" the inspector may recommend some type of road furniture at a specific location, even if it is not planned. An indication of the need for such a recommendation may be the frequent occurrence of serious accidents at a particular location.

Culverts, Small Structures and Bridges

The existence and location of these structures as indicated on the strip map, mostly transverse hydrological structures, should be mentioned on the inventory of the sections to be inspected (left side of Form III -- Figure III.3). The functional condition (blockage or erosion) and the structural condition (collapsing or cracking) of these structures must be inspected. They are evaluated on a three point scale, described in the damage catalogue.

Damage to be Repaired Urgently

Carriageway damage which jeopardises the safety of the road user or the flow of traffic (see also Damage Catalogue) should be repaired immediately. This damage (or damages) is recorded in the section "Remarks" by recording the damage code (see Annex C) and its (their) location.

III.4. IMPLEMENTATION

Road condition surveys should be carried out over the whole network where possible. If resources are limited, the surveys should be conducted on the most important roads first, in general those with highest traffic.

III.4.1. Frequency

Road condition surveys should be carried out preferably at least once per year and most frequently on important roads. On less important roads the frequency may be less, depending also on normal maintenance procedures. For example, if, as a routine maintenance procedure, a road is dragged or graded twice a year, less frequent monitoring may be required. If maintenance planning procedures are in operation or are about to be implemented, it is possible to carry out the RCS just once.

In areas with pronounced rainy seasons it is recommended that road condition surveys be undertaken before the start of the rainy season, enabling the maintenance units to perform preventive maintenance measures. A second survey should be performed at the end of the rainy season to make an inventory of damage and urgently required repairs.

III.4.2. Preparations before the survey

Before beginning the survey, the inspector in charge will complete the general information block of the form including the identification of the inspection team, the date and the weather conditions. It is assumed that the regional office has completed this form for the road sections to be inspected.

The forms for the various sections included in the survey for the day will be arranged in the planned sequence of the survey itinerary.

III.4.3. The two phases of the survey

Each survey will be carried out in two phases:

First Phase

The complete daily road itinerary is driven over, the speed being in accordance with the design speed (as defined in Annex B), without violating the regulations in force or the safety of other road users. The mechanical integrity of the vehicle should also be considered when travelling at high speed.

This first survey phase permits the inspection team to evaluate the roughness (evenness) of the carriageway according to the evaluation method described in Annex B.

Recording the time required to drive through a section will also be useful for the evaluation of the carriageway condition on the basis of average vehicle speed.

Second Phase

The second survey phase will be performed at a low speed (e.g. 15 km/h) to record the other characteristics. The inspector identifies the different sections by following the itinerary sheet and records the characteristics of the section by completing the inspection form section by section.

At least every 500 m a stop is made for a physical inspection and to record the specific data of the section just inspected.

Additional stops have to be made at every structure indicated on the itinerary scheme and at every other place of particular interest.

It is useful to take photographs of particular problems where it is difficult to give a numerical value to the characteristic concerned. These photographs can be of great help later at the decision making stage.

III.5. ASSESSMENT OF RESULTS

The interpretation of the road condition survey results and computerized data analysis are dealt with in Chapter V. The present sub-chapter only deals with the summarising of the survey results and their interpretation to determine the possible need for a further detailed visual inspection.

The survey system developed in this Chapter ensures the objective and integral assessment of road network conditions. Hereunder, a suggested method for using the survey data is developed, but the user of this manual is free to use these data as he sees fitting; the data can be utilised in many different ways to suit the needs of the user (see Chapter V).

III.5.1. Summary of survey results

To facilitate the interpretation of the collected data, the results of the itineraries are grouped into tables. The layout of these forms is such that updating the data bank is easy. There are two forms, one for the carriageway and one for the road structures (Forms IV and V). When Form II is received at the office the average values per section are calculated and entered on Form IV. Form V gives the condition value for each structure on the basis of the individual Form III's received.

Apart from the data relevant to each section (columns 3 to 8 on Form IV), the annual average daily traffic (AADT) and the road identification are recorded in columns 1, 2 and 9 (as defined in Chapter II).

III.5.2. Interpretation of survey results

For a given section the interpretation of the survey results can lead to one of the following conclusions (to be recorded in column 14 on Form IV):

a) The section in question is in a satisfactory condition. No intervention is needed in the short or medium term. Normal maintenance activities should continue as planned.

b) The road condition is such that a maintenance strategy modification is required, or that a maintenance or rehabilitation scheme is required in the medium term.

   If this need is on the carriageway (Column 11 of Form IV, Class 3), a detailed visual inspection is to be carried out (see Chapter IV) to define the required maintenance and repair works. If the need is dictated by damage to the "other" elements or to the small structures and bridges, maintenance activities should be planned without making a detailed visual inspection. These activities can be planned in the medium term if the value accorded to the damage is not higher than two.

Figure III.4. **SUMMARY OF ROAD CONDITION SURVEY RESULTS: AVERAGE CONDITION VALUE ("FORM IV")**

ROAD CONDITION SURVEY (RCS) PAVED AND UNPAVED ROADS

OFFICE:	DISTRICT:		PAGE
Established by:			
Date:			

DESIGNATION OF ROAD SECTION								
Road		Section No.	Identification Points (IP)			Section length	Pavement width	AADT
Class	No.		I.P.	Designation	Km	Km	m	Veh/day
1	2	3	4	5	6	7	8	9
			From To		From To			
			From To		From To			
			From To		From To			
			From To		From To			
			From To		From To			
			From To		From To			
			From To		From To			
			From To		From To			
			From To		From To			
			From To		From To			

SURVEY DATA			
Avg. condition value		Road signs & Furniture	Notes / Remarks
Carriageway	Road side elements		
11	12	13	14

Figure III.5. **SUMMARY OF SURVEY RESULTS FOR STRUCTURES: AVERAGE CONDITION VALUE PER STRUCTURE ("FORM V")**

ROAD CONDITION SURVEY (RCS)

OFFICE:	DISTRICT:	PAGE
Established by:		
Date:		

STRUCTURES				CONDITION SURVEY DATA: AVERAGE CONDITION VALUE			
Road denomination/Class:							
Type of structure	Main section	Sub-section	Chainage	Silted/ Blocked	Scour	Structural damage	Notes / Remarks

c) Certain repairs must be undertaken immediately to avoid an unusable or extremely dangerous road. This is especially true for severe but not extensive damage to the carriageway and for damage related to small structures and bridges valued at 3.

Urgent maintenance activities (c above) must be carried out immediately. The need for immediate intervention may be requested by the inspection team by radio. Immediate action may be necessary when a road in critical condition is encountered and no time should be wasted in reporting this.

Chapter IV

DETAILED VISUAL INSPECTION

IV.1. PURPOSE

The purpose of the Detailed Visual Inspection (DVI) is to record the type, extent and severity of damage. It helps the engineer to determine the causes of damage and identify the appropriate repair measures. The DVI is conducted only on those sections which have been identified from the road condition survey results as being in a suspect condition.

The relation of Detailed Visual Inspection to other elements in the monitoring scheme is shown in Figure IV.1.

IV.2. METHOD OF EVALUATION

Each detailed visual inspection parameter is to be assessed on a five-point scale. Conditions are rated by:

a) The extent of the damage, expressed as the percentage of the surface area and/or

b) The severity of the damage.

Most condition recording systems use this two dimensional evaluation system.

IV.3. INSPECTION PROCEDURES

IV.3.1. Forms

Form I (Figure II.5) is required to prepare the daily itinerary for the inspection team. At this stage, a clear distinction must be made between paved and unpaved roads. Forms VI (Figure IV.2) for unpaved roads, and VII (Figure IV.3) for paved roads, are applicable for the detailed visual inspection.

Figure IV.1. **POSITION OF DETAILED VISUAL INSPECTION IN THE GENERAL SCHEME**

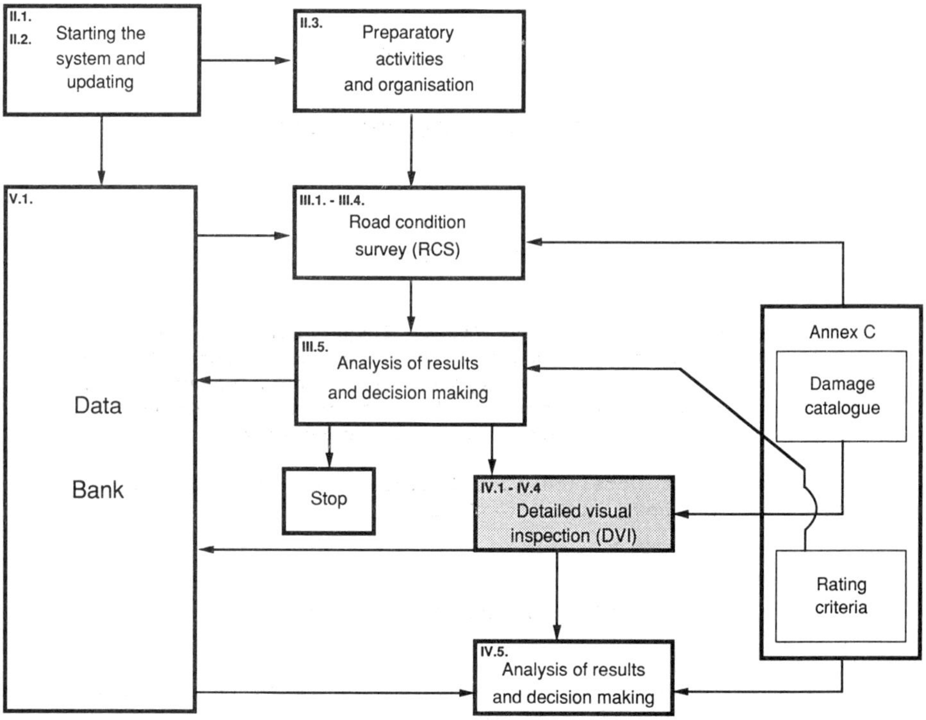

It should be noted that for both unpaved and paved roads other types of damage may be defined. In order to keep the forms "workable" only the most common types of damage are included.

IV.3.2. Parameters for unpaved roads

At this level of inspection the following, most common, parameters can be distinguished (Form VI -- Figure IV.2):

-- Rutting;
-- Corrugations;
-- Camber/Crossfall;
-- Gravel thickness;
-- Erosion gullies;
-- Potholes;
-- Clay.

Figure IV.2. **DETAILED VISUAL INSPECTION FORM FOR UNPAVED ROADS ("FORM VI")**

DETAILED VISUAL INSPECTION (DVI)				**UNPAVED ROADS**	
OFFICE:	DISTRICT:		ROAD DENOMINATION / CLASS:		Main Section No.
Name of inspector:			FROM I.P.:		Km:
Date:	Weather o Clear o Rainy	Carr.way: o Dry o Drying o Wet	TO I.P.:		Km:
			Pavement type:	Pavement width: m	Section length: Km

SUB-SECTION		○						○						○						○						○					
Roadside		L			R			L			R			L			R			L			R			L			R		
Damage type - Cat	E \ S	L	M	S	L	M	S	L	M	S	L	M	S	L	M	S	L	M	S	L	M	S	L	M	S	L	M	S	L	M	S
Rutting - UR1	<10%	1	3	5	1	3	5	1	3	5	1	3	5	1	3	5	1	3	5	1	3	5	1	3	5	1	3	5	1	3	5
	10-50%	2	4	5	2	4	5	2	4	5	2	4	5	2	4	5	2	4	5	2	4	5	2	4	5	2	4	5	2	4	5
	>50%	3	5	5	3	5	5	3	5	5	3	5	5	3	5	5	3	5	5	3	5	5	3	5	5	3	5	5	3	5	5
Corrugations - UR2	<10%	1	3	4	1	3	4	1	3	4	1	3	4	1	3	4	1	3	4	1	3	4	1	3	4	1	3	4	1	3	4
	10-50%	2	3	5	2	3	5	2	3	5	2	3	5	2	3	5	2	3	5	2	3	5	2	3	5	2	3	5	2	3	5
	>50%	3	4	5	3	4	5	3	4	5	3	4	5	3	4	5	3	4	5	3	4	5	3	4	5	3	4	5	3	4	5
Loss of camber - UR3	<10%	1	3	4	1	3	4	1	3	4	1	3	4	1	3	4	1	3	4	1	3	4	1	3	4	1	3	4	1	3	4
	10-50%	2	3	5	2	3	5	2	3	5	2	3	5	2	3	5	2	3	5	2	3	5	2	3	5	2	3	5	2	3	5
	>50%	3	4	5	3	4	5	3	4	5	3	4	5	3	4	5	3	4	5	3	4	5	3	4	5	3	4	5	3	4	5
Gravel thickness - UR4	<10%	1	3	5	1	3	5	1	3	5	1	3	5	1	3	5	1	3	5	1	3	5	1	3	5	1	3	5	1	3	5
	10-50%	2	3	5	2	3	5	2	3	5	2	3	5	2	3	5	2	3	5	2	3	5	2	3	5	2	3	5	2	3	5
	>50%	3	4	5	3	4	5	3	4	5	3	4	5	3	4	5	3	4	5	3	4	5	3	4	5	3	4	5	3	4	5
Erosion gullies - UR5	<10%	1	3	5	1	3	5	1	3	5	1	3	5	1	3	5	1	3	5	1	3	5	1	3	5	1	3	5	1	3	5
	10-50%	2	4	5	2	4	5	2	4	5	2	4	5	2	4	5	2	4	5	2	4	5	2	4	5	2	4	5	2	4	5
	>50%	3	5	5	3	5	5	3	5	5	3	5	5	3	5	5	3	5	5	3	5	5	3	5	5	3	5	5	3	5	5
Potholes - UR6	<5 (No./100m)	1	3	5	1	3	5	1	3	5	1	3	5	1	3	5	1	3	5	1	3	5	1	3	5	1	3	5	1	3	5
	5-15	2	4	5	2	4	5	2	4	5	2	4	5	2	4	5	2	4	5	2	4	5	2	4	5	2	4	5	2	4	5
	> 15	3	5	5	3	5	5	3	5	5	3	5	5	3	5	5	3	5	5	3	5	5	3	5	5	3	5	5	3	5	5
Clay - UR7	< 5%	1			1			1			1			1			1			1			1			1			1		
	5-50%	2			2			2			2			2			2			2			2			2			2		
	>50%	3			3			3			3			3			3			3			3			3			3		
....																															

Remarks: ____

Figure IV.3. **DETAILED VISUAL INSPECTION FORM FOR PAVED ROADS ("FORM VII")**

DETAILED VISUAL INSPECTION (DVI)				**PAVED ROADS**
OFFICE:	DISTRICT:	ROAD DENOMINATION/ CLASS:		Main Section No.
Name of inspector:		FROM I.P.:		Km:
Date:	Weather o Clear o Rainy / Carr.way: o Dry o Drying o Wet	TO I.P.:		Km:
		Pavement type:	Pavement width: m	Section length: Km

SUB-SECTION		○						○						○						○						○					
Roadside		L			R			L			R			L			R			L			R			L			R		
Damage type - Cat	E \ S	L	M	S	L	M	S	L	M	S	L	M	S	L	M	S	L	M	S	L	M	S	L	M	S	L	M	S	L	M	S
Rutting - PR1	<10%	1	3	5	1	3	5	1	3	5	1	3	5	1	3	5	1	3	5	1	3	5	1	3	5	1	3	5	1	3	5
	10-50%	2	4	5	2	4	5	2	4	5	2	4	5	2	4	5	2	4	5	2	4	5	2	4	5	2	4	5	2	4	5
	>50%	3	5	5	3	5	5	3	5	5	3	5	5	3	5	5	3	5	5	3	5	5	3	5	5	3	5	5	3	5	5
Corrugations - PR2	<10%	1	3	4	1	3	4	1	3	4	1	3	4	1	3	4	1	3	4	1	3	4	1	3	4	1	3	4	1	3	4
	10-50%	2	3	5	2	3	5	2	3	5	2	3	5	2	3	5	2	3	5	2	3	5	2	3	5	2	3	5	2	3	5
	>50%	3	4	5	3	4	5	3	4	5	3	4	5	3	4	5	3	4	5	3	4	5	3	4	5	3	4	5	3	4	5
Depressions - PR3	<10%	1	3	4	1	3	4	1	3	4	1	3	4	1	3	4	1	3	4	1	3	4	1	3	4	1	3	4	1	3	4
	10-50%	2	4	5	2	4	5	2	4	5	2	4	5	2	4	5	2	4	5	2	4	5	2	4	5	2	4	5	2	4	5
	>50%	3	5	5	3	5	5	3	5	5	3	5	5	3	5	5	3	5	5	3	5	5	3	5	5	3	5	5	3	5	5
Transverse cracking - PR4a	<2 (No./100m)	1	3	4	1	3	4	1	3	4	1	3	4	1	3	4	1	3	4	1	3	4	1	3	4	1	3	4	1	3	4
	2-15	2	3	5	2	3	5	2	3	5	2	3	5	2	3	5	2	3	5	2	3	5	2	3	5	2	3	5	2	3	5
	>15	3	4	5	3	4	5	3	4	5	3	4	5	3	4	5	3	4	5	3	4	5	3	4	5	3	4	5	3	4	5
Longitudinal cracking - PR4b	<10%	1	3	5	1	3	5	1	3	5	1	3	5	1	3	5	1	3	5	1	3	5	1	3	5	1	3	5	1	3	5
	10-50%	2	4	5	2	4	5	2	4	5	2	4	5	2	4	5	2	4	5	2	4	5	2	4	5	2	4	5	2	4	5
	>50%	3	5	5	3	5	5	3	5	5	3	5	5	3	5	5	3	5	5	3	5	5	3	5	5	3	5	5	3	5	5
Alligator cracking - PR4c	<10%	1	4	5	1	4	5	1	4	5	1	4	5	1	4	5	1	4	5	1	4	5	1	4	5	1	4	5	1	4	5
	10-50%	2	5	5	2	5	5	2	5	5	2	5	5	2	5	5	2	5	5	2	5	5	2	5	5	2	5	5	2	5	5
	>50%	3	5	5	3	5	5	3	5	5	3	5	5	3	5	5	3	5	5	3	5	5	3	5	5	3	5	5	3	5	5
Holes - PR5	<5 (No./100m)	1	3	5	1	3	5	1	3	5	1	3	5	1	3	5	1	3	5	1	3	5	1	3	5	1	3	5	1	3	5
	5-15	3	4	5	3	4	5	3	4	5	3	4	5	3	4	5	3	4	5	3	4	5	3	4	5	3	4	5	3	4	5
	> 15	4	5	5	4	5	5	4	5	5	4	5	5	4	5	5	4	5	5	4	5	5	4	5	5	4	5	5	4	5	5
Edge distress - PR6	<10%	1	3	4	1	3	4	1	3	4	1	3	4	1	3	4	1	3	4	1	3	4	1	3	4	1	3	4	1	3	4
	10-50%	2	3	5	2	3	5	2	3	5	2	3	5	2	3	5	2	3	5	2	3	5	2	3	5	2	3	5	2	3	5
	>50%	3	4	5	3	4	5	3	4	5	3	4	5	3	4	5	3	4	5	3	4	5	3	4	5	3	4	5	3	4	5
Stripping/ Fretting/ Ravelling - PR7a	<10%	1	2	4	1	2	4	1	2	4	1	2	4	1	2	4	1	2	4	1	2	4	1	2	4	1	2	4	1	2	4
	10-50%	1	3	5	1	3	5	1	3	5	1	3	5	1	3	5	1	3	5	1	3	5	1	3	5	1	3	5	1	3	5
	>50%	2	5	5	2	5	5	2	5	5	2	5	5	2	5	5	2	5	5	2	5	5	2	5	5	2	5	5	2	5	5
Stripping/Fretting/ Ravelling of surface - PR7b	<15%	1	2	3	1	2	3	1	2	3	1	2	3	1	2	3	1	2	3	1	2	3	1	2	3	1	2	3	1	2	3
	15-30%	2	4	5	2	4	5	2	4	5	2	4	5	2	4	5	2	4	5	2	4	5	2	4	5	2	4	5	2	4	5
	>30%	3	5	5	3	5	5	3	5	5	3	5	5	3	5	5	3	5	5	3	5	5	3	5	5	3	5	5	3	5	5
Bleeding - PR8	< 5%	1			1			1			1			1			1			1			1			1			1		
	5-50%	2			2			2			2			2			2			2			2			2			2		
	>50%	3			3			3			3			3			3			3			3			3			3		

Remarks: ______

It is possible to add other parameters which may differ from country to country.

The detailed visual inspection will be carried out on the road surface or carriageway.

The parameters will be assessed in relation to their influence on vehicle operating costs, traffic speed and driving safety. The actual "measuring device" is given in the damage catalogue.

IV.3.3. Parameters for paved roads

The parameters to be evaluated are (Form VII -- Figure IV.3):

-- Rutting;
-- Corrugations;
-- Depressions;
-- Cracking: Longitudinal; Transverse;
-- Alligator cracking;
-- Potholes;
-- Edge distress;
-- Stripping/Fretting/Ravelling;
-- Bleeding.

For simplicity some damage types may be combined; for example "longitudinal and transverse cracking", which are two forms of linear cracking. Alligator cracking is a well advanced stage of interconnected "cracking". If combinations of damage are encountered, e.g. rutting with cracking, both types of damage have to be recorded on the same inspection form.

The parameters will be assessed in relation to their influence on vehicle operating costs, traffic speed and driving safety. The actual "measuring device" is given in the Damage Catalogue.

IV.4. IMPLEMENTATION

IV.4.1. Frequency

Detailed visual inspections do not have to be carried out regularly at a specified frequency. Only when the road condition survey indicates that the road condition is not satisfactory, is a detailed visual inspection required.

IV.4.2. Preparations before the inspection

Before starting the inspection, the inspector in charge completes the general information block on the inspection forms. The forms for the various sections included in the itinerary for the day will be arranged in the planned sequence of the inspection itinerary.

IV.4.3. Method

The inspection is carried out by walking the sub-section to be inspected. The inspector uses the Damage Catalogue and the proposed inspection method to judge the different parameters. For each type of damage the value to be recorded is based on the relevant matrix given on the damage sheet.

Where required, the measurements are performed with chain, straight edge or other available equipment. Often it is necessary to make some notes to permit a realistic evaluation of the whole sub-section. After the inspection of the complete sub-section the inspector completes the form.

IV.5. ASSESSMENT OF RESULTS

IV.5.1. Summary of inspection results

On both forms the average values of the sub-sections have to be combined into an arithmetic length weighted average for the complete section. These average figures will be carried to Form VIII (Figure IV.4), to update the data bank.

IV.5.2. Interpretation of inspection results

For a given road section the interpretation of the inspection results may lead to one of the following conclusions:

a) A maintenance or rehabilitation scheme is required in the short term.

b) A maintenance or rehabilitation scheme is required in the medium term.

It is assumed that all necessary urgent repairs are carried out immediately after the road condition survey.

Figure IV.4. **SUMMARY OF DETAILED VISUAL INSPECTION RESULTS: AVERAGE CONDITION VALUE ("FORM VIII")**

DETAILED VISUAL INSPECTION (DVI)

OFFICE:	DISTRICT:	PAGE
Established by:		
Date:		

DESIGNATION OF ROAD SECTION								
Road		Section / Sub-section No.	Identification Points (IP)			Section length	Carriageway width	AADT
Class	No.		I.P.	Designation	Km	Km	m	Veh/day
1	2	3	4	5	6	7	8	9
			From		From			
			To		To			
			From		From			
			To		To			
			From		From			
			To		To			
			From		From			
			To		To			
			From		From			
			To		To			
			From		From			
			To		To			
			From		From			
			To		To			
			From		From			
			To		To			
			From		From			
			To		To			
			From		From			
			To		To			

	Visual inspection data																	Avg.cond.value	Instrumental survey data		
	Unpaved roads								Paved roads												
Roadside	Rutting	Corrugations	Loss of camber	Gravel thickness	Erosion gullies	Potholes	Clay		Rutting	Corrugations	Depressions	Tr/Long cracking	Alligator crack.	Holes	Edge distress	Stripping	Bleeding		IRI	Deflection	Skid resistance
11	12	13	14	15	16	17	18	19	20	21	22	23	24	25	26	27	28	29	30	31	32
L																					
R																					
L																					
R																					
L																					
R																					
L																					
R																					
L																					
R																					
L																					
R																					
L																					
R																					
L																					
R																					
L																					
R																					
L																					
R																					

Chapter V

DATA AGGREGATION AND ANALYSIS

V.1. DATA PROCESSING AND STORAGE

V.1.1. Validating the data

Whatever method of data acquisition is used, the accuracy of the data provided by road surveys must be checked before they are put to any use.

Dependent upon the type of data involved, checking should be done either by the unit that performed the inspections, or at a management level that is in close contact with the field. Inspection forms or route diagrams summarising the inspection results may provide an excellent medium for the validation process, provided they are systematically used both in the field and at management level.

This "manual" validation process may be completed by the application of plausibility checks, whether elementary, requiring a given value to be within a specified range, or more complex (prohibiting the simultaneous occurrence of specified conditions.

Another requirement for effective data validation is that all the data collected must be entered in a common referencing system.

A typical example might be as follows. The distance between two position marks is known to be 2,500 m; this distance was established during the first survey of the road system, and is henceforth deemed to be correct. However, in a second survey, either to measure a different feature or to update existing measurements, the distance between the position marks comes to 2 550 m. All distances measured to the position mark concerned can now be corrected to the common base of 2,500 m, by a processing operation based on a simple rule of three.

Such inaccuracy in measurements of distance is normal; it is caused by inaccuracy in the measuring apparatus itself and by the response of the operator when passing position marks. Ideally, therefore, the distance between two successive position marks should not be greater than 5 km.

V.1.2. Filing the data

In some organisations, data may simply be filed in filing cabinets, containing for example the completed route diagrams or inspection forms. Storage may be arranged on a route or area basis. Where such a filing system is used there should be no need to make fair copies of forms completed on site: this wastes time and involves the risk of errors when information is transcribed.

The limitations of this type of data filing are that updating and analysis can be cumbersome and slow, particularly where the road network is very extensive.

In most cases, the data should be stored in a digital form on a computerised database. As there is always a need to summarise and report the data, and perform some kinds of analyses, the usefulness of computer systems becomes very apparent. Microcomputer systems are now within the financial and operational capabilities of most highway agencies, and at least the central office of an agency should store and process their data on computers. The original, "raw" data should be stored on hardcopy (paper files) as backup to a computer system, and, in district offices for some countries, may be the only form of data summary. For large road networks, microcomputer systems may have inadequate capacity, and linkages to mini- or mainframe computers may need to be established.

Some requirements for building a computerised data bank are given below.

a) All information in the data bank must be accurately defined and coded. A Data Dictionary, clearly defining the various headings (methods of collection, updating and quantification) and containing the coding system, must always be established beforehand.

b) The responsibilities of the different staff involved in compiling and maintaining the data bank, and also the validation procedures, must be clearly specified. In particular, there must be clear rules indicating who may access data for consultation and who may access them for amendment purposes.

c) From the data processing point of view, the data bank will be composed of separate files, with consistency provided by the referencing system. Updating will be easier if each file contains few headings, but operations such as data linkage will be more complicated if the data concerned belong to several different files. A compromise must therefore be made.

d) The Data Base Management System (DBMS) used for compiling the data bank and for accessing and processing the data must be selected on the criteria of speed of access to information (with allowance for the expected volume of data and available hardware) and ease of operation. Using a DBMS, even one considered usable by a non-specialist, requires a certain amount of training in its use.

e) Using micro-computers reduces the cost of compiling, maintaining and accessing the data base. Also, the software is usually easier for non-specialists to use.

The efficiency of a computer system in handling a database depends on the size of the database files, and on the processing speed and memory capacity of the computer. Microcomputers of:

-- Medium level (8 MHz processing speed and 40 Mb hard disk capacity, e.g., IBM, PC, AT); and

-- High level (more than 20 MHz processing speed and hard disk capacity more than 100 Mb, e.g. 386-chip machines);

have been used successfully for road networks of up to 50 000 km length. Lower levels of computer are generally not recommended unless the database is very small, e.g. less than 10 Mb.

Whenever a data bank is established it is essential that precise specifications are compiled beforehand, particularly with regard to the referencing system and the operations envisaged (some operations may be extremely difficult if the data are not well-organised).

V.1.3. Main types of data processing

Basic processing of the information contained in the data bank is of three very different types:

-- Technical processing involving the preparation of overall analyses to facilitate work scheduling;

-- Statistical processing involving the production of various specific statistics relating to one or more types of data (histograms, correlations, regressions);

-- Data aggregation.

More complex processing is described in Section V.2.

Technical Processing

The most common type of processing is the generation of route diagrams which summarise all the data on a given subject, e.g. traffic operation, road safety, road maintenance, etc.

These synoptic diagrams are used to assist in work scheduling, but they are not usually sufficient for precise specification of the work to be done.

Figure V.1 provides an example of a route diagram extracted from Gabon's Road Data Bank.

For maintenance or strengthening works, these diagrams are used to determine the locations and types of additional testing (boreholes, deflections, radius of deflection curvature, etc.) required in order to accurately specify the type of work and/or treatment needed.

Statistical Processing

Methods of statistical processing are too numerous for exhaustive description here; this review will therefore be limited to some selected treatments and a few examples.

a) Item counts and histograms

This type of operation involves the simple arithmetic processing of data stored under various headings, in some cases after entering one or two qualifiers.

For example:

-- The number of traffic signs (a simple count); or

-- The total length of road with surface dressing (a count involving arithmetic operations on the data).

The only difficulty here is how to define the input query precisely enough. Even a question as simple as the surface area of a road may give rise to ambiguity: is it the area between edge markings, where these exist, or the total paved area? Should surfaced shoulders be included when calculating surface areas?

The results of single-qualifier counts can be presented simply in the form of tables, figures, histograms, "pie" charts, etc. Figure V.2 is an example of a histogram using selected pavement criteria drawn automatically (Source: Gabon Road Data Bank).

b) Correlations and regressions

The organisation in a computerised data bank of data collected for a road network is a powerful tool for developing technical knowledge about pavement performance.

In particular, the study of trends in pavement behaviour and condition, relative to such factors as weather, traffic and structural design, makes it possible:

-- To acquire an objective view of the quality of methods used in designing new pavements and strengthening or maintaining existing ones; and

-- To analyse alternative maintenance strategies on the basis of a life cycle analysis of maintenance costs.

Studying these trends will also help to determine the most cost-effective frequency for updating the various parameters in the data base.

Data Aggregation

A global index of condition that combines the values of various condition parameters, is often desired to represent the overall condition of a road or the network. While such a summary index can be useful for reporting

Figure V.1. **EXAMPLE OF ROUTE DIAGRAMME (GABON)**

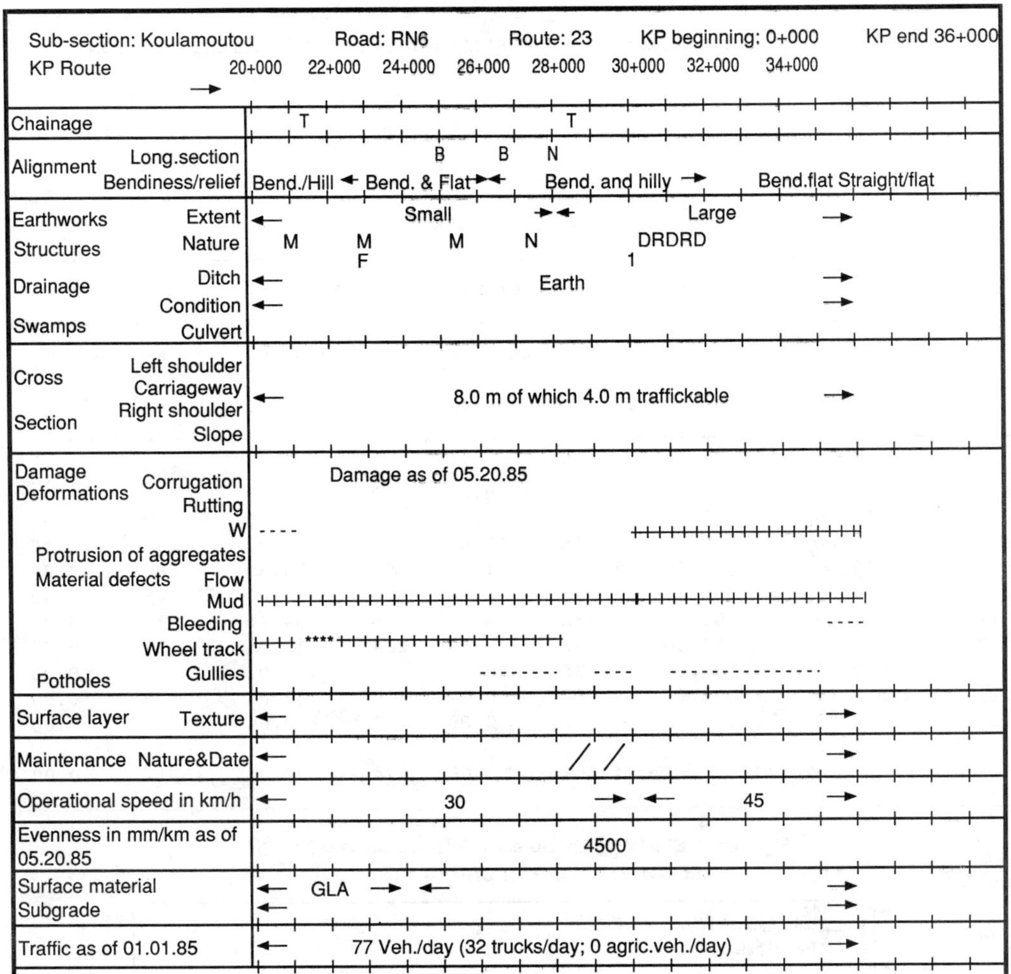

Legend:

Chainage	C = Cross junction T = Cross-town link A = Quarry	G = T-junction to left K = Milestone B = Rain barrier	D = T-junction to right E = Waterway F = Railway	P = Police (post) R = Airport I = Miscellaneous
Alignment-Long. section	G = Left bend D = Right bend	H = Summit B = Sag	+ = Uphill - = Downhill	
Earthworks (nature)	N = Natural soil	R = Embankment	D = Cutting	M = Mixed
Structures Bridge types	F = Forest depot 1 = OAB | 2=OAME	B = Pit | D = Culvert 3 = OABA | 4 = OABP	M = Retaining wall 5 = OAMI | 6 = OACF	G = Hut
Damage	Extent of	---- From 0 to 10%	++++From 10 to 50%	*****More than 50%
Maintenance Paved Unpaved	ED = Dense mix LO = Localised	BB = Bitum. concrete RP = Reshaping	ES = Surface dress. RC = Regravelling	RN = Strengthening

Figure V.2

EXAMPLE OF SUMMARY HISTOGRAM USING SELECTED PAVEMENT CRITERIA (GABON)

HISTOGRAM : TRAFIC traffic/day   Date: 23/04/86

CONCERNS

TOTAL NETWORK

Name of histogram: HISTOMJA

Class	Limits		Total		Paved		Unpaved	
	DE (incl.)	A (exclu.)	L1 (Km)	L2 (%)	L3 (Km)	L4 (%)	L5 (km)	L6 (%)
1	0.00	1.00	460	10.48	11	2.32	449	11.46
2	1.00	10.00	0	0.00	0	0.00	0	0.00
3	10.00	25.00	0	0.00	0	0.00	0	0.00
4	25.00	50.00	818	18.63	0	0.00	818	20.88
5	50.00	100.00	1 398	31.84	94	19.83	1 304	33.29
6	100.00	250.00	995	22.66	0	0.00	995	25.40
7	250.00	500.00	432	9.84	115	24.26	317	8.09
8	500.00	1 000.00	238	5.42	204	43.04	34	0.87
9	1 000.00	2 500.00	38	0.87	38	8.02	0	0.00
10	2 500.00	5 000.00	12	0.27	12	2.53	0	0.00
11	5 000.00	9 999.00	0	0.00	0	0.00	0	0.00
	Total		4 391	0.00	474	0.00	3 917	0.00

Figure V.2. **EXAMPLE OF SUMMARY HISTOGRAMME USING SELECTED PAVEMENT CRITERIA (GABON)**

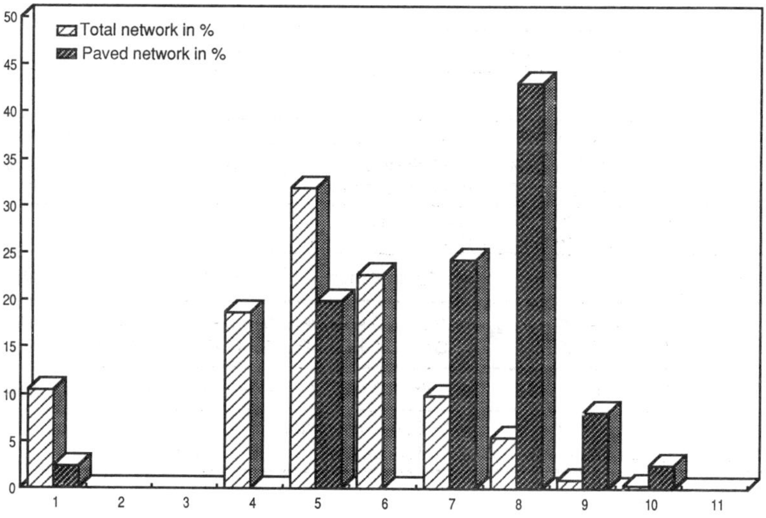

at a network level, it has been found inadequate for scheduling maintenance works because it is inadequate to distinguish between functional and structural, or corrective and preventive, needs.

As a means of overcoming this type of drawback, it has often been found preferable to perform an analysis based on the values of individual parameters. As a minimum, it is best to report condition in at least two parameters, namely roughness, which represents the impact on user costs and riding quality, and surface distress, which indicates the need for maintenance to prevent escalation of the roughness in the future. This suggests that data aggregation cannot be envisaged without having considered the goals pursued.

In the framework of this manual, three distinct levels of aggregation can be selected:

-- That required by decision-makers who have a use for general indicators by which to monitor pavement quality, mainly in order to assess the service provided or, in some cases, the value of the existing infrastructure;

-- That required by a centralised agency that needs to assess long-term maintenance requirements, to draw up general work schedules and allocate funds where they are needed, and to rank work proposals by order of priority or urgency;

-- That required by field engineers who have to produce detailed work plans.

Data aggregation suitable to these three levels are discussed in turn in the following section.

V.2. USE OF DATA FOR EVALUATING PAVEMENT CONDITION AND WORK PROGRAMMES

V.2.1. <u>Overall monitoring of pavement condition</u>

The most simple aggregate rating of condition is into three or more categories such as good, fair and poor. But to be meaningful, these categories need to be defined by physical levels of roughness and surface distress, and to be related to traffic levels, or service standards.

A general rating which assesses the condition in relation to the level of service to the user, based on a weighted combination of the gathered data, may be developed. The North American Present Serviceability Index (PSI) which correlated panel ratings to condition for paved roads, is not considered appropriate for general use because the rating reference and the implicit weighting of condition are not transferable to different countries and environments. Roughness is the foremost aspect of condition affecting the level of service to the user and the economic benefits to be derived from maintenance, and it is preferable to relate this to a transferable physical scale, such as the International Roughness Index (IRI) (see Annex B). As noted above, it is also desirable to have a separate measure of surface distress because this can indicate the potential for accelerated deterioration in the future.

Other methods of data aggregation may be considered so as to be able to respond to specific technical problems encountered in the road network to be assessed. For example:

-- The low level of service provided by a system of unpaved roads might be quantified by highly aggregated data, such as travel speed or vehicle path regularity;

-- Conversely, in the case of major paved road systems, which generally have good roughness and transverse profile characteristics, other features, such as surface texture or the lateral friction coefficient of the surfacing, might be prime factors in defining level of service.

Other parameters not directly related to pavement surface condition may influence the definition of service provided. In particular, the trafficability of a road will be an essential factor; this may be restricted because of bad weather, by load restrictions or mud patches caused by periods of rain.

It is clearly impossible to separate the level of service from pavement quality which determines the need for load restrictions during or after bad weather, according to the bearing capacity of the subsoil, for instance.

The data used in rating pavements are sometimes limited to data from the visual inspection; however, it is often thought desirable to include other parameters that are more representative of structural behaviour, such as deflection or pavement design features (type and thickness of pavement layers).

Whatever the means of data aggregation used, there are two accepted methods of presenting the aggregate data:

-- The first is to show the ratings for a given year on a map, so as to depict pavements in relation to their condition; this method simply shows routes with different levels of service, and differences from one geographic area to another;

-- The second method involves plotting histograms, bar charts or "pie" charts showing the percentages of roads providing good or poor levels of service; these are used mainly to compare results at different points in time, to show good or bad trends in levels of service.

Figure V.3 is an example of how pavement condition can be followed up.

V.2.2. Specifying the jobs needed

Detailed work proposals should never be based solely on computerised data aggregation.

Figure V.3. **EXAMPLE OF OVERALL MONITORING RESULT OF PAVEMENT CONDITIONS**

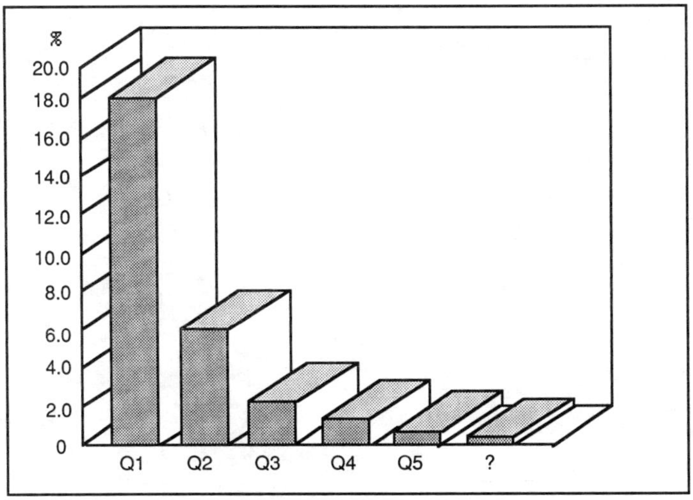

Legend:
Q1 = Good
Q2 = Acceptable
Q3 = Unsatisfactory
Q4 = Bad
Q5 = Very bad
? = Investigation needed

The maintenance engineer should first analyse the various quantified and unquantified sets of source data, to try to understand the cause of the trouble and establish a diagnosis.

To diagnose the causes of distress and to evaluate remedies, it may be necessary to conduct specific tests after a detailed inspection of the section to be treated. These may include, inter alia, structural analysis of the pavement (thickness, nature and quality of the different courses), an assessment of the subgrade by means of mechanical tests (e.g. CBR), deflection measurements etc.

Proceeding from the diagnosis phase to the specification of remedial works often involves unorthodox aggregation of available data, a screening process in accordance with the engineer's experience which will add decisive weight to certain of the factors involved.

Although this approach obviously cannot be codified, it is nevertheless advisable to have manuals or guidelines that partially specify the process and offer typical solutions; these may be in the form of tables, for example, in which case the table entry procedure must be flexible enough to ensure that experience is still assigned its proper role.

Chapter VI

TRAINING

VI.1. GENERAL

Training is of critical importance.

The successful use of this manual and the effective and efficient implementation of the road monitoring scheme depend on appropriate training and retraining. Training forms an integral part of the monitoring system; it must be planned together with the necessary budgetary provisions. The costs involved may be substantial but the funds required constitute a good investment for ensuring maximum benefits of road monitoring and maintenance management.

The main steps in the development of a training programme are:

-- Identification of training needs;

-- Training needs analysis;

-- Planning of the training programme;

-- Execution of the training programme;

-- Evaluation of the results.

Within the scope of road monitoring systems the key personnel are the district engineer and the road maintenance engineer/road inspector. The training programmes need to be designed for these personnel to:

-- Enhance their recognition of the need for improved maintenance procedures;

-- Develop a positive attitude toward the monitoring system;

-- Improve technical ability and skills.

VI.1.1. Identification of training needs

The identification of training needs depends largely on a review of the work or job description. The job description identifies the following components:

-- Quality required;
-- Quantity required;
-- Need for communication;
-- Responsibilities.

VI.1.2. Training needs analysis

Having recognised the need for training in the maintenance organisation, a detailed training survey is mandatory to ascertain the training requirements and to develop a training programme.

After a close study of the structure of the organisation where training is required, the key personnel will be the subject of a job analysis.

By analysing the job and matching the job holder's ability to this analysis, the individual training needs can be assessed and compiled to establish the total training needs of sections, departments or of the organisation as a whole.

VI.1.3. Planning of the training programme

Based on the survey of training needs, training courses or modules can be developed for individuals or groups of staff. Countless combinations of training methods are possible so that in each case the training specialists can devise the training package most suited to the operational requirements.

The careful timing of all training programmes is essential to provide the proper implementation of road monitoring activities.

The choice of training methods, trainers and training materials must always be carefully considered, bearing in mind the cost effectiveness of the training programmes, together with the availability of facilities, equipment and training aids.

VI.1.4. Execution of the training programme

The execution of the training programme -- if the preceding phase of training planning has been undertaken effectively -- will be a relatively straight forward operation.

However, a few components of the training should be given special attention:

-- The operational requirements of the organisation or projects should not be disrupted by taking the trainees away from their jobs for unacceptably long periods;

-- The training methods should be carefully selected, e.g. classroom lectures, coaching, tutorials, overseas training, etc.;

-- Good examples of road maps with sections and sub-sections have to be available;

-- Some representative road sections have to be selected before the actual training starts;

-- Presentation methods should be selected;

-- Intermediate evaluation of trainees and trainers should be undertaken to provide continuous feedback and allow for minor modifications of the training presentation.

VI.1.5. Evaluation of the results

No training project is complete without a final evaluation of its effectiveness in terms of quality or performance of the trainees, the trainer and the programme as a whole.

During the actual training the performance of the trainees is observed and this observation continues when the trainees resume their normal duties.

The performance of the trainees is compared with the training objectives and any variation can lead to additional advice or coaching. The evaluation also covers the quality of the training given, the trainer and the training service itself, with the aim of providing the best possible service.

VI.2. JOB DESCRIPTION

The job descriptions given under this item only deal with those elements of the functions that are related to the road monitoring system. A complete job description of the key functions is beyond the scope of this publication.

VI.2.1. Job description of the district engineer

Basic functions:

-- Organisation of the road monitoring system in his district;

-- Processing of the road monitoring information into multi-year and annual road maintenance programmes;

-- Advising the central authorities about the technical and financial implications of the road monitoring results.

Organisational relationship:

-- Accountable to the central authorities for the proper fulfillment of his functions;

-- Supervises road inspectors.

Principal activities:

-- Prepares, organises and processes periodic road inspections, assesses and reports the results and establishes annual and multi-year maintenance programmes;

-- Prepares, organises and processes periodic traffic counts; assesses and reports the results;

-- Co-ordinates the work of the road inspection teams;

-- Updates the records of road sections and sub-sections.

VI.2.2. Job description of the road inspector

Basic functions:

-- Day-to-day field operations in accordance with the established road inspection programme;

-- Advising the district engineer on the needs and progress of his field activities.

Organisational relationship:

-- Accountable to the district engineer;

-- Supervises the inspection teams.

Principal activities:

-- Monitors the condition of the road sections under his responsibility;

-- Prepares and submits to the district engineer weekly and monthly operating records.

VI.3. OUTLINE FOR TRAINING COURSES

A procedure for training courses can be proposed for:

-- District engineers (DE);

-- Road inspectors (RI).

The starting point is the design of separate training programmes for the DE and the RI.

The programme for the DE will mainly deal with subjects such as:

-- Data gathering and storage using the inspection forms;

-- Data interpretation and processing for the design of a maintenance programme.

The programme for the RI will mainly cover issues such as:

-- Familiarisation with the inspection forms and their use in the field;

-- Familiarisation with the Damage Catalogue and the inspection manual.

Of course there is some overlap between these training courses, since the DE should be familiar with the operational constraints of surveying, the forms, damage catalogue and inspection manual. The RI will be able to perform his task with additional motivation if he is aware of some of the procedures used in the processing of the forms and to what conclusions they may lead. Therefore, the DE should participate in the training of the RI and the RI should receive some general information about the processing of the forms.

VI.3.1. Training of the DE

The training aims at familiarising the DE with the following items:

-- The road monitoring system;

-- The gathering of fixed and variable data;

-- The processing of these data to develop a detailed maintenance programme for the short and medium term;

-- The classification of road sections and sub-sections for inspection and maintenance purposes;

-- The organisation of traffic counts and processing of the results;

-- The use and function of the damage catalogue and the inspection manual;

-- Assessment of the road inspectors and their function.

VI.3.2. Training of the RI

The training aims at familiarising the RI with the following items:

-- The need for surveying road sections in general;

-- The structure, layout and use of the road condition survey form;

-- The structure, layout and use of the detailed inspection form;

-- The use of the inspection manual and the damage catalogue;

-- How to complete the forms and deal with potential problems;

-- The use of the equipment required for the road condition survey and detailed visual inspection;

-- The importance of the uniformity of inspection results.

VI.4. TRAINEE SELECTION

The selection of trainees depends largely upon the organisation of the road department in a particular country. In the procedures described above the district engineer and the road inspector were designated as the key personnel. This is a logical distribution of responsibilities, but other possibilities exist.

Training depends on the local organisation of the road department and it is likely that the introduction of a road monitoring system will be the first step toward the implementation of a nationwide road maintenance organisation where this is not in existence.

VI.5. TRAINING LOCATION

Training sessions must be in classrooms for the theoretical aspects, but an important part of the training has to be given in the field. It is the trainer's responsibility to ensure that enough "field material" will be available for the training sessions. The training must familiarise the trainees with driving over various road surfaces at varying speeds to enable the trainees to give a uniform rating. It is important that the trainee sees and appreciates these differences.

For the detailed visual inspection it is important that the trainees learn to accurately judge the condition of the road elements.

VI.6. CONCLUSIONS

In the appendix some possible training packages are presented. These packages are for guidance only since the design of a scientific training programme depends on the road department's organisation.

When introducing a road monitoring system, it is critically important that the optimal functioning of the system is assured. In most cases, the assistance of professionals is required.

These professionals should have an excellent knowledge of the way a road department should be organised and the function of the road monitoring system.

It is also important that a nucleus of local trainers be trained to ensure a continuation of training for the local road maintenance staff.

APPENDIX TO CHAPTER VI

Possible training packages

The following illustrates possible training packages which may be further developed for key personnel, mentioned in Section VI.1. and for managers involved in the road maintenance system.

Training package 1

Title	: Road monitoring system
Target group/function	: Management level
Entry level	: Management level
Methods	: 10 days-classroom instruction 5 days-exercises
Materials and aids	: - maps - manuals - handouts
Instructor	: Expert
Duration	: 3 weeks
Participants	: District engineers and managers
Location	: Central office and the field

Training package 2

Title	: Road network inventory
Target group/function	: District engineers
Entry level	: Middle grade engineer
Methods	: 2 days-classroom instruction 2 days-exercises 5 days-post-course performance evaluation
Materials and aids	: - maps - manuals - handouts - drawing equipment
Instructor	: Expert
Duration	: 2 weeks
Participants	: 2 per district
Location	: Central office and the field

Training package 3

Title	: Road condition survey
Target group/function	: Road inspectors and district engineer
Entry level	: Mid-level technicians
Methods	: 2 days-classroom instruction
	: 2 days-exercises
	: 5 days-post-course performance evaluation
Materials and aids	: - road network inventory
	- manuals
	- handouts
	- equipment
Instructor	: Expert
Duration	: 2 weeks
Participants	: 2-4 per district
Location	: Central office and the field

Training package 4

Title	: Detailed visual inspection
Target group/function	: Road inspectors and district engineer
Entry level	: Mid-level technicians
Methods	: 5 days-classroom instruction
	: 5 days-exercises
	: 10 days-post-course performance evaluation
Materials and aids	: - road network inventory
	- manuals
	- handouts
	- equipment
Instructor	: Expert
Duration	: 4 weeks
Participants	: 2-4 per district
Location	: Central office and the field

Training package 5

Title	: Traffic surveying
Target group/function	: Road inspectors and district engineer
Entry level	: Mid-level technicians
Methods	: 5 days-classroom instruction
	: 5 days-post-course performance evaluation
Materials and aids	- manuals
	- handouts
	- equipment to determine axle loads
Instructor	: Expert
Duration	: 2 weeks
Participants	: 2-4 per district
Location	: Central office and the field

Training package 6

Title	: Road maintenance data management
Target group/function	: District engineers
Entry level	: Middle grade engineer
Methods	: 5 days-classroom instruction
	: 5 days-post-course performance evaluation
Materials and aids	- manuals
	- handouts
	- computer software (where appropriate)
Instructor	: Expert
Duration	: 2 weeks
Participants	: 1 per district
Location	: Central office

Training package 7

Title	: Training of local trainers
Target group/function	: All relevant functions
Entry level	: As required for subject matters
Subjects and modules	: To be defined/developed
Methods	: 2 weeks-classroom instruction and exercises
	: 1 week in-service guidance
Materials and aids	- manuals
	- handouts
	- closed circuit television (CCTV)
Instructor	: Consultant
Duration	: 3 weeks
Participants	: Several groups of 10 persons each
Location	: Central office

These training packages are an outline only. The actual training needs can only be determined by experts in road monitoring systems and training who assess the existing skills and design the training packages and the required materials and aids, accordingly.

Annex A

BASIC INSTRUCTIONS FOR IMPLEMENTING INSPECTION ACTIVITIES

The purpose of the inspections is to collect and update the information on all roads to give a complete record of the network. The overall use of the various forms is presented in Figure A.1.

A.1. ITINERARY

The itinerary is prepared by the responsible department using:

-- A road map of the itinerary (Figures A.2, A.3 and A.4).

-- A strip map of the sections with an indication of special elements such as structures and culverts (Figure A.5).

-- Form I which gives in tabular form a list of sections and sub-sections, their references and identifications (Figure A.6).

A.2. ROAD CONDITION SURVEY (RCS)

The condition survey sheets are prepared and arranged in sequence:

-- Form II for the carriageway, road side elements, road signs and furniture (see Figure A.7) records the condition of the various elements to be inspected by sub-section and allows for the calculation of the average condition per section for the carriageway and road side elements.

-- Form III (see Figure A.8) lists on the left side the bridges, culverts, etc. to be inspected, in the sequence in which they are encountered along the section(s), together with the section number (and sub-section) and the chainage. On the right side of the form the condition of the various elements to be inspected is recorded.

Figure A.1. **SUMMARY PRESENTATION OF USE OF FORMS**

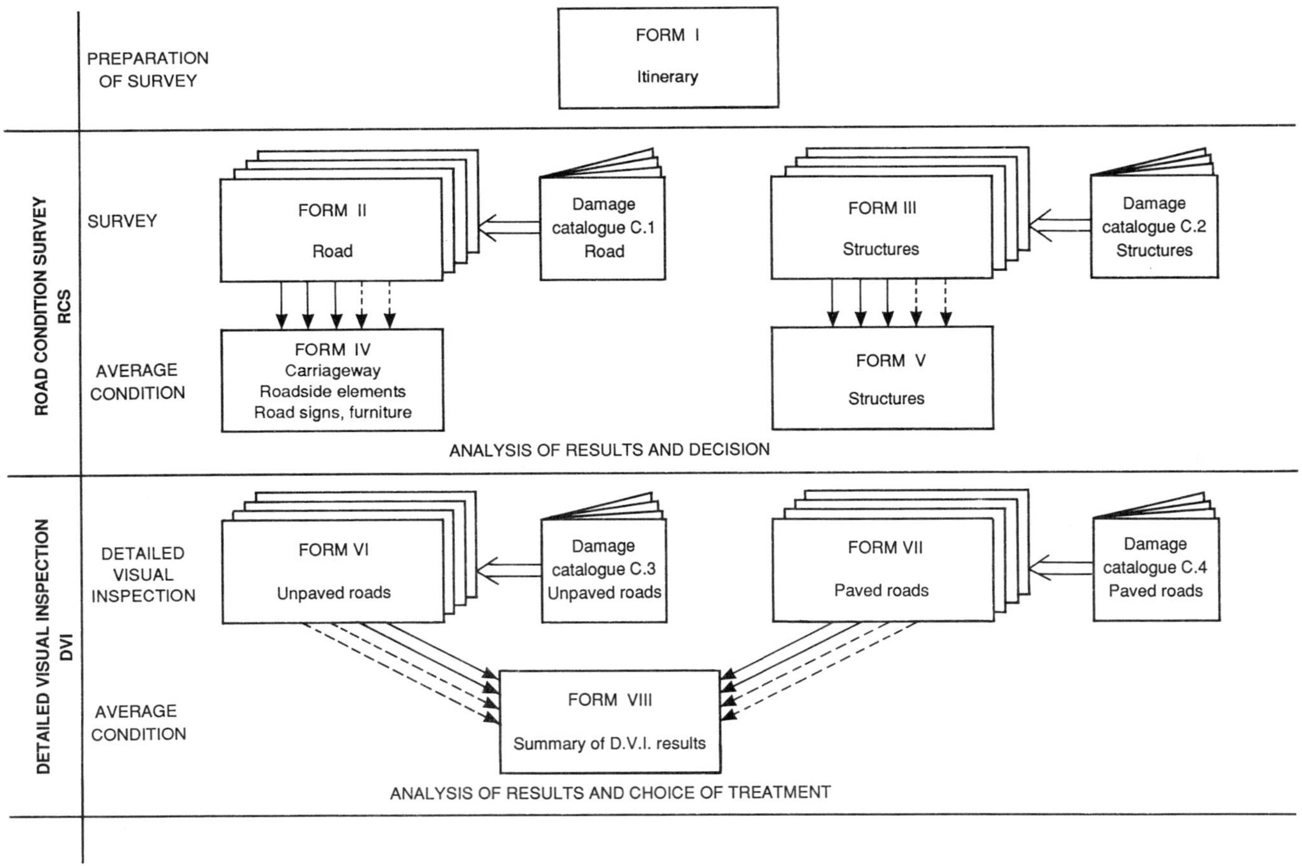

Figure A.2. **TOPOGRAPHICAL MAP OF MOROKULI DISTRICT**
(SCALE 1: 500 000)

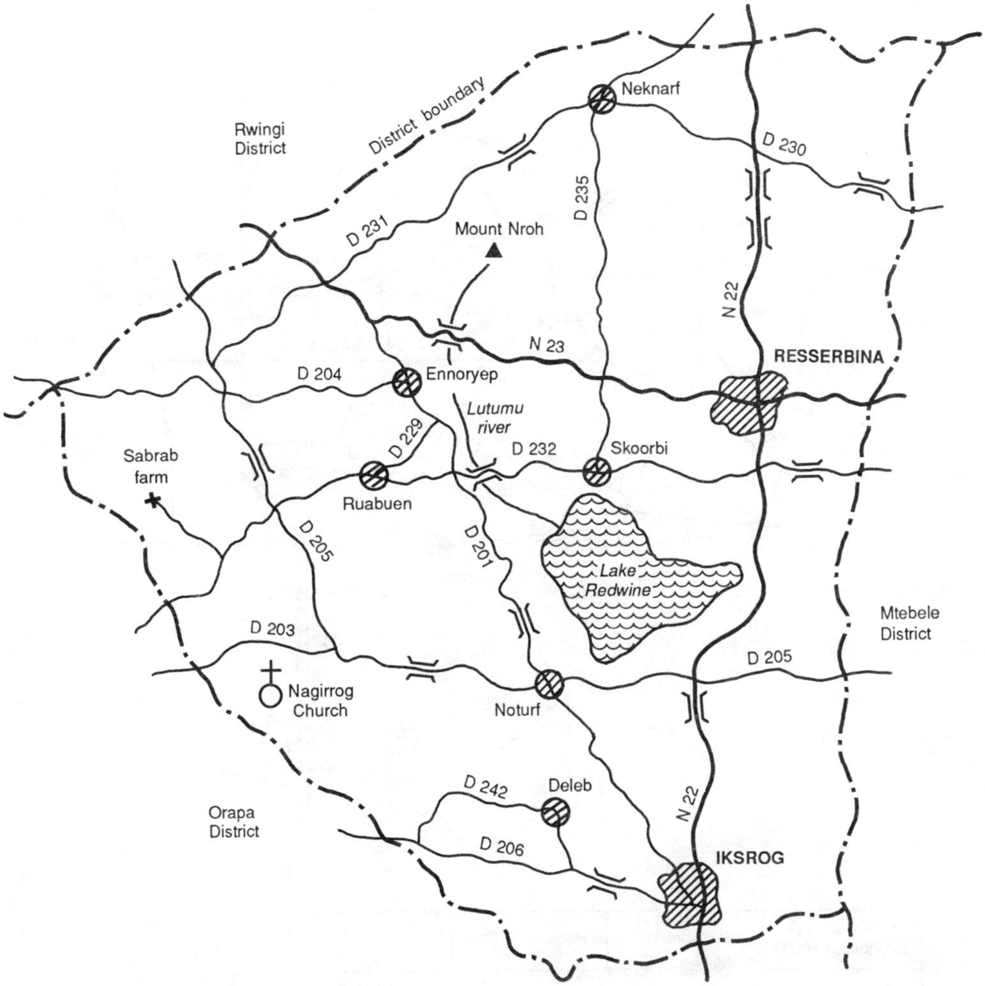

Figure A.3. **SIMPLIFIED ROAD MAP OF MOROKULI MAINTENANCE DISTRICT**

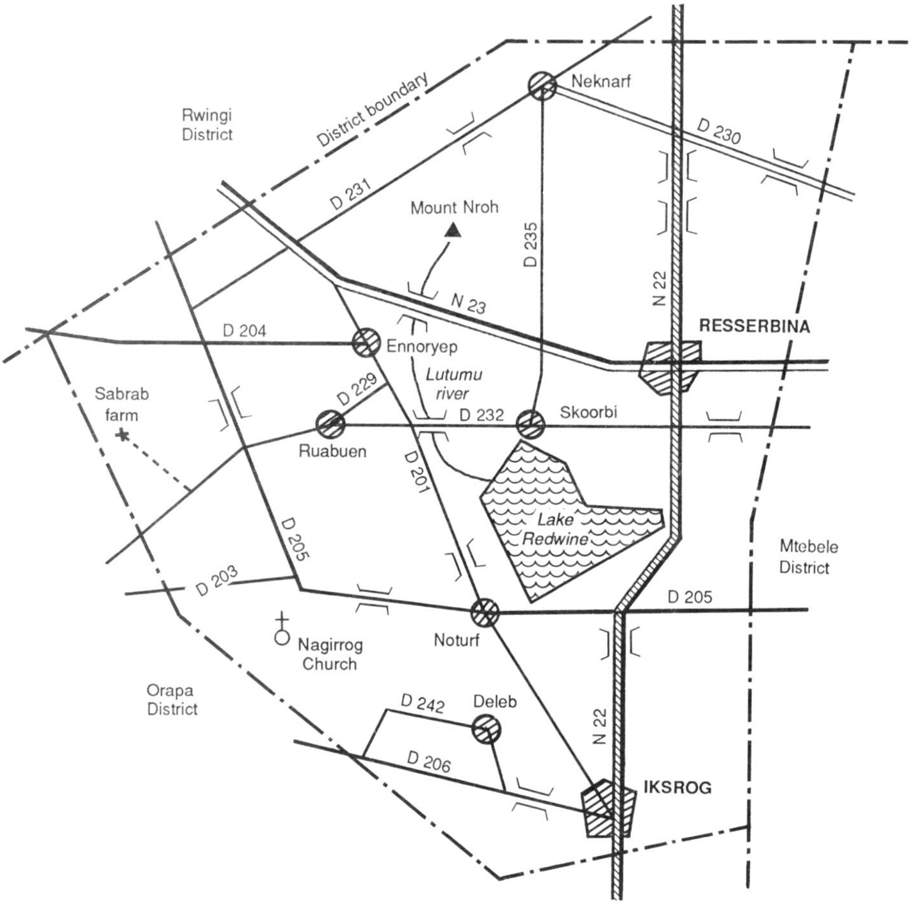

KEY

Annual average daily traffic (AADT)	Surface type	Road width (m.)	
> 1000	bituminous	7.0 -7.5	
200 - 1000	bituminous	5.5 - 7.0	
< 200	bituminous	5.0 - 5.5	
> 200	gravel	7.0	
50 - 200	gravel	5.0 - 7.0	
< 50	gravel	3.5 - 5.0	

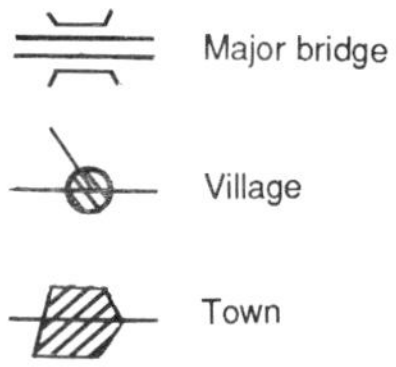

Figure A.4. **SIMPLIFIED ROAD MAP OF MOROKULI MAINTENANCE DISTRICT WITH SECTIONS**

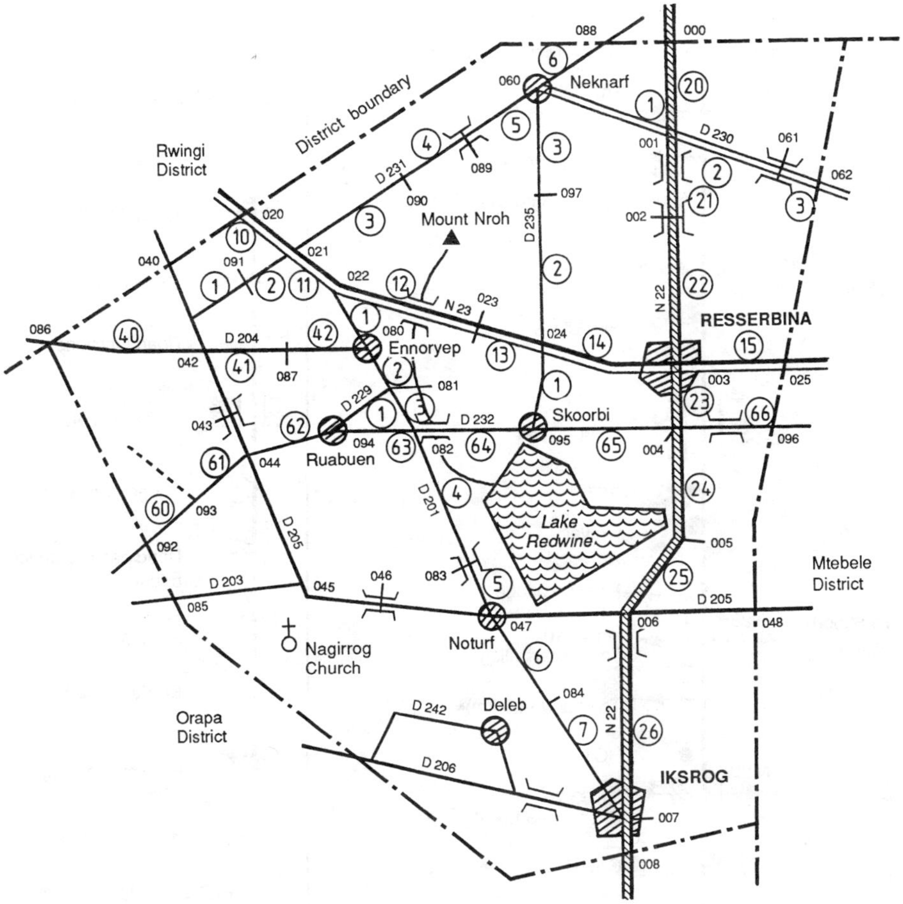

KEY

Annual average daily traffic (AADT)	Surface type	Road width (m.)	
> 1000	bituminous	7.0 -7.5	
200 - 1000	bituminous	5.5 - 7.0	
< 200	bituminous	5.0 - 5.5	
> 200	gravel	7.0	
50 - 200	gravel	5.0 - 7.0	
< 50	gravel	3.5 - 5.0	-----

Major bridge

Village

Town

Figure A.5. **STRIP MAP OF ROAD N23**

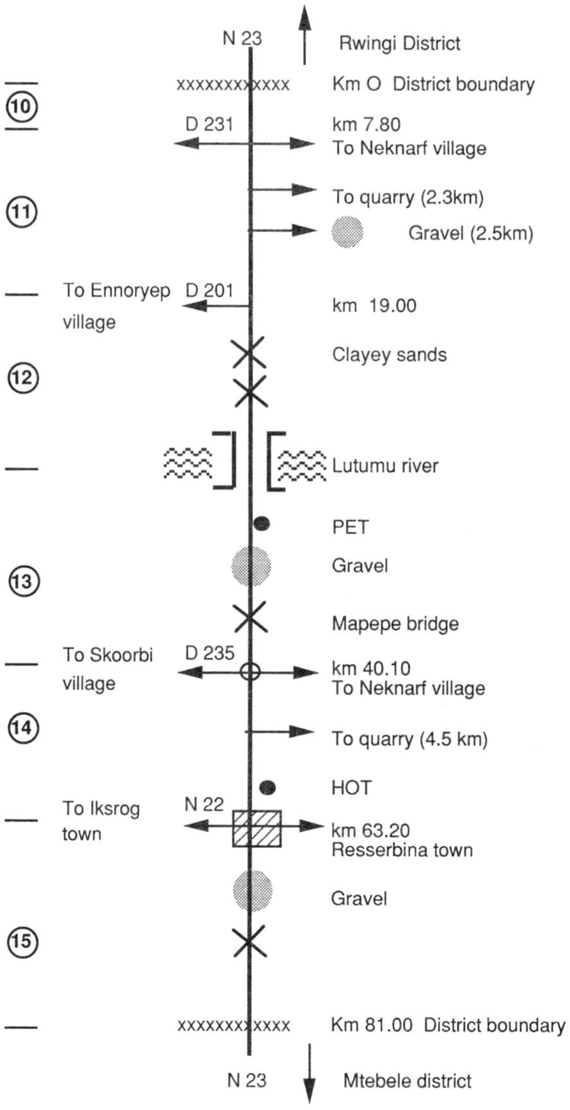

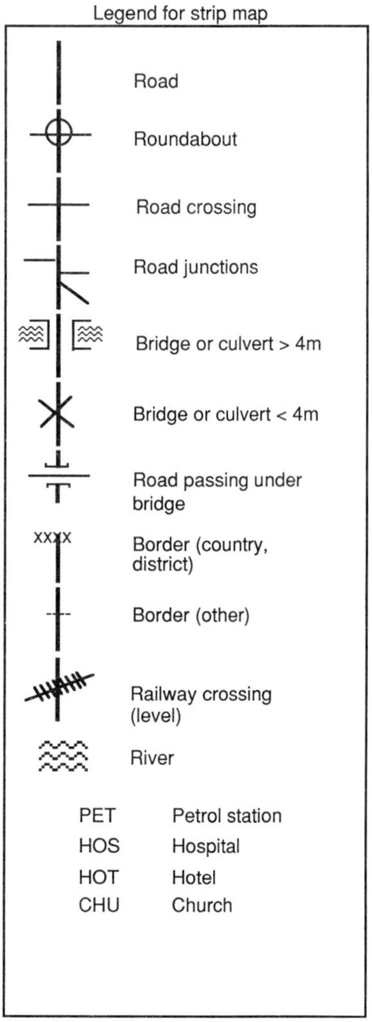

Figure A6. **FORM FOR SURVEY (RCS) OR INSPECTION (DVI) ITINERARY "FORM I"**

MAINTENANCE DISTRICT: MOROKULI Completed by: A. MUTALI	Itinerary No.: 2	PAGE (1)

ROUTE: Identification points (I.P.) Kind	Designation	MAIN SECTIONS: Road Class	Road No.	No. of main section	Chainage (km)	Length Single (km)	Length Added (km)
1	2	3	4	5	6	7	8
V	080/Ennoryep	D	201	(1)	223.50	7.05	
J	022				216.45		7.05
J	022	N	23	(11)	216.45	5.50	
C	021				210.95		12.55
C	021	N	23	(10)	210.95	5.95	
L	020				205.00		18.50
C	021	D	231	(3)	12.00	10.50	
CT	090				22.50		29.00
CT	090	D	231	(4)	22.50	7.25	
B	089				29.75		36.25
B	089	D	231	(5)	29.75	8.45	
V	060				38.20		44.70
V	060	D	231	(6)	38.20	9.05	
L	088				47.25		53.75
V	060	D	235	(3)	33.35	10.35	
CC	097				23.00		64.10
CC	097	D	235	(2)	23.00	14.90	
C	024				8.10		79.00
C	024	D	235	(1)	8.10	8.10	
J	095				0.00		87.10

Sub-sections: Number of sub-sections	(1) From	(1) To / (2) From	(2) To / (3) From	(3) To / (4) From	(4) To / (5) From	(5) To	No. of main section
IP	V	CV	CC	CP	J		1
Station	223.50	222.00	218.55	217.35	216.45		
Length (km)	1.50	3.45	1.20	0.90			
IP	J	B	C				11
Station	216.45	214.35	210.95				
Length (km)	2.10	3.40					
IP	C	CV	CV	L			10
Station	210.95	209.70	207.85	205.00			
Length (km)	1.25	1.85	2.85				
IP	C	CV	CV	B	CV	CT	3
Station	12.00	14.50	15.25	18.50	20.15	22.50	
Length (km)	2.50	0.75	3.25	1.65	2.35		
IP	CT	CV	CV	CC	B		4
Station	22.50	23.65	24.85	26.50	29.75		
Length (km)	1.15	1.20	1.65	3.25			
IP	B	CC	CV	V			5
Station	29.75	34.45	36.65	38.20			
Length (km)	4.70	2.20	1.55				
IP	V	CV	CC	CV	CV	L	6
Station	38.20	39.65	41.85	44.55	46.35	47.25	
Length (km)	1.45	2.20	2.70	1.80	0.90		
IP	V	CT	CV	CV	CV	CC	3
Station	33.35	30.15	28.45	26.95	24.65	23.00	
Length (km)	3.20	1.70	1.50	2.30	1.65		
IP	CC	B	CV	V	V	C	2
Station	23.00	19.95	18.85	14.15	9.65	8.10	
Length (km)	3.05	1.10	4.70	4.50	1.55		
IP	C	B	CV	J			1
Station	8.10	5.35	4.85	0.00			
Length (km)	2.75	0.50	4.85				

Figure A.6.(cont'd.) **WORK FORM FOR SURVEY (RCS) OR INSPECTION (DVI) ITINERARY "FORM I"**

MAINTENANCE DISTRICT: MOROKULI Completed by: A. MUTALI	Itinerary No.: 2	PAGE 2

ROUTE		MAIN SECTIONS					
Identification points (I.P.)		Road		No. of main section	Chai-nage (km)	Length	
Kind	Designation	Class	No.			Single (km)	Added (km)
1	2	3	4	5	6	7	8
C	024	N	23	13	237.55	6.50	87.10
CP	023				231.05		93.60
CP	023	N	23	12	231.05	14.60	
J	022				216.45		108.20

Sub-sections							
Number of sub-sections	① From	① To / ② From	② To / ③ From	③ To / ④ From	④ To / ⑤ From	⑤ To	No. of main section
IP	C	CV	CV	CP			13
Station	237.55	234.95	233.65	231.05			
Length (km)	2.60	1.30	2.60				
IP	CP	CV	B	CV	J		12
Station	231.05	227.95	226.35	224.15	216.45		
Length (km)	3.10	1.60	220	7.70			
IP							
Station							
Length (km)							
IP							
Station							
Length (km)							
IP							
Station							
Length (km)							
IP							
Station							
Length (km)							
IP							
Station							
Length (km)							
IP							
Station							
Length (km)							
IP							
Station							
Length (km)							
IP							
Station							
Length (km)							

Figure A.7. **ROAD CONDITION SURVEY (RCS) FORM ("FORM II")**

ROAD CONDITION SURVEY PAVED AND UNPAVED ROADS		ROAD DENOMINATION / CLASS: D 201		Main Section No.
		FROM: 080 (Ennoryep)		Km: 223.50
		TO: 022		Km: 216.45
OFFICE: R.MAINT	DISTRICT: MOROKULI	Pavement type: DST	Pavement width: 6.50 m	Section length: 7.05 Km
Name of inspector: J. Mbanaw		Summary (Avg. Condition)	Carriageway: 25 : B = 3.125	
Date: 88 0120	Weather ■ Clear o Rainy; Carr.way: o Dry ■ Drying o Wet		Road side elements: (2.2 + 1.8 + 2.2 + 1.5) / 4=1.9	
			Remarks:	

SUB-SECTION	①		②		③		④		⑤	
Chainage of sub-section	223.50		222.00		218.55		217.35		216.45	
Roadside	L	R	L	R	L	R	L	R	L	R
Carriageway	1 2 ④ 3 5	1 ② 4 3 5	1 2 ④ 3 5	1 2 4 ③ 5	1 2 ④ 3 5	1 ② 4 3 5	1 2 ④ 3 5	1 ② 4 3 5	1 2 4 3 5	1 2 4 3 5
Prevailing damage	CL	PO	CA	PO	CA	CL	CA	CL		

ROAD SIDE ELEMENTS	Damage type	L	R	L	R	L	R	L	R	L	R
Shoulder	Deformation RCS1	1 ② 3	1 2 ③	① 2 3	1 2 ③	1 ② 3	1 2 ③	① 2 3	1 ② 3	1 2 3	1 2 3
	Scour RCS2	1 ② 3	1 ② 3	1 ② 3	1 ② 3	① 2 3	1 2 ③	① 2 3	1 ② 3	1 2 3	1 2 3
Side drains	Siltation RCS3	1 ② 3	1 2 ③	① 2 3	1 2 ③	① 2 3	1 2 ③	1 ② 3	① 2 3	1 2 3	1 2 3
	Scour RCS4	1 ② 3	1 ② 3	① 2 3	1 2 ③	1 ② 3	1 2 ③	1 ② 3	① 2 3	1 2 3	1 2 3
Debris/Vegetation Encroachment RCS5		① 2 3	1 ② 3	① 2 3	1 2 ③	1 ② 3	1 2 ③	1 ② 3	1 ② 3	1 2 3	1 2 3
Obstacles/ Obstructions RCS6		1 ② 3	1 2 ③	① 2 3	① 2 3	1 ② 3	① 2 3	① 2 3	① 2 3	1 2 3	1 2 3
		1 2 3	1 2 3	1 2 3	1 2 3	1 2 3	1 2 3	1 2 3	1 2 3	1 2 3	1 2 3
		1 2 3	1 2 3	1 2 3	1 2 3	1 2 3	1 2 3	1 2 3	1 2 3	1 2 3	1 2 3
Average Condition (Others)		11 + 15 = 26 26 : 12 = 2.2		7 + 15 = 22 22 : 12 = 1.8		10 + 16 = 26 26 : 12 = 2.2		9 + 9 = 18 18 : 12 = 1.5		+ = : =	

ROADSIGNS/FURNITURE											
Chainage		223.05	219.95	221.35	215.55	216.15	216.45				
		L	R	L	R	L	R				
Dirty	RCS7	1 ② 3	① 2 3	① 2 3	① 2 3	1 ② 3	1 ② 3	1 2 3	1 2 3	1 2 3	1 2 3
Damaged	RCS7	1 ② 3	1 ② 3	① 2 3	1 ② 3	① 2 3	1 ② 3	1 2 3	1 2 3	1 2 3	1 2 3
Missing	RCS7	① 2 3	① 2 3	① 2 3	① 2 3	1 ② 3	1 2 ③	1 2 3	1 2 3	1 2 3	1 2 3
		1 2 3	1 2 3	1 2 3	1 2 3	1 2 3	1 2 3	1 2 3	1 2 3	1 2 3	1 2 3

REMARKS: 1) D.V.I. required (average > 3.0) for this section
2) Road furniture at 216.45 to be repaired / replaced

Figure A.8. **CONDITION SURVEY FORM FOR STRUCTURES ("FORM III")**

ROAD CONDITION SURVEY (RCS)

OFFICE: R.D. MAINT.	DISTRICT: MOROKULI	PAGE 1
Name of inspector: J.Mbanaw		
Date: 88 01 20	Weather ☒ Clear o Rainy	

STRUCTURES

Road denomination/Class:

CONDITION SURVEY

(Circled rating shown in parentheses.)

Type of structure	Main section	Sub-section	Chainage	Silted/Blocked L	Silted/Blocked R	Scour L	Scour R	Structural Damage L	Structural Damage R	Notes/Remarks
Culvert ∅ 200cm	1	.1	222.00	(1) 2 3	1 (2) 3	1 2 (3)	1 (2) 3	(1) 2 3	(1) 2 3	
Culvert ∅ 80 cm	1	.3	218.25	1 (2) 3	1 (2) 3	(1) 2 3	(1) 2 3	(1) 2 3	(1) 2 3	
Culvert ∅ 80 cm	1	.3	218.15	1 (2) 3	1 (2) 3	(1) 2 3	(1) 2 3	(1) 2 3	(1) 2 3	
Culvert ∅ 80 cm	1	.4	217.15	(1) 2 3	1 (2) 3	1 (2) 3	1 (2) 3	(1) 2 3	(1) 2 3	
Bridge 5 m span	1	.4	216.50	(1) 2 3	(1) 2 3	1 (2) 3	(1) 2 3	(1) 2 3	1 2 (3)	road kerbs to be replaced
				1 2 3	1 2 3	1 2 3	1 2 3	1 2 3	1 2 3	
				1 2 3	1 2 3	1 2 3	1 2 3	1 2 3	1 2 3	
				1 2 3	1 2 3	1 2 3	1 2 3	1 2 3	1 2 3	
				1 2 3	1 2 3	1 2 3	1 2 3	1 2 3	1 2 3	
				1 2 3	1 2 3	1 2 3	1 2 3	1 2 3	1 2 3	
				1 2 3	1 2 3	1 2 3	1 2 3	1 2 3	1 2 3	
				1 2 3	1 2 3	1 2 3	1 2 3	1 2 3	1 2 3	
				1 2 3	1 2 3	1 2 3	1 2 3	1 2 3	1 2 3	

A.2.1. Parameters to be evaluated in RCS

Group	Type	No.	No. of rating levels	Section*
Carriageway	Surface quality	-	5	Annex B
	Prevailing damage (if necessary)	-	-	3-4
Shoulders	Deformation	RCS 1	3	1
	Scour	RCS 2	3	
Side drains	Siltation	RCS 3	3	1
	Scour	RCS 4	3	
Road space	Debris/vegetation encroachment	RCS 5	3	1
	Obstacles/Obstructions	RCS 6	3	
Road signs & furniture; road markings	Dirty and/or damaged, missing	RCS 7	3	1
Culverts	Silted/blocked	RCS 8	3	2
	Scour	RCS 9	3	
	Structural damage	RCS 10	3	
Bridges	Silted/blocked	RCS 11	3	2
	Scour	RCS 12	3	
	Structural damage	RCS 13	3	
Damage & defects to be repaired immediately	Various, to be mentioned under "Remarks"	-	-	3-4

* Numbers refer to Sections of the Damage Catalogue, published separately.

A.2.2. Survey procedure

Two phases:

1. First phase

To evaluate the quality of the road carriageway according to the method described in Annex B, drive without stopping through the complete section. The carriageway is evaluated using the average speed and the comfort during the run and the average level of drainage.

2. Second phase

To collect the other characteristics as mentioned under A.2.1. (Form II), drive through the section at a much lower speed (approx. 15 km/h). A stop is scheduled at least every 500 m for close inspection of shoulders and ditches (Form II, see Figure A.7), and also at all the structures, culverts and bridges (Form III, see Figure A.8), indicated on the strip map. The condition of the elements is evaluated according to rating criteria stated in the Damage Sheets of the Catalogue (Sections 1 and 2). The situation and damage to be repaired urgently are indicated on the form under "Remarks".

A.2.3. Summary of results

The average condition is calculated from the scores for a) carriageway and b) roadside elements, then recorded on Form IV (Figure A.9). When computing the average carriageway score, one point for every prevailing damage recording of CA (alligator cracking) or PO (potholing) should be added to the sum before dividing by the number of subsection-wheelpaths.

A.3. DETAILED VISUAL INSPECTION (DVI)

A more detailed visual inspection is carried out on those sections that were suspected to be in a critical condition as determined by the road condition survey. The isolated sections should be regrouped into a daily itinerary for further inspection. During this inspection the parameters to be evaluated differ for paved and unpaved roads.

A.3.1. Parameters for DVI of unpaved roads, Form VI

Type	No.	Code	Function affected	Points in scale
Rutting	DVI/UR 1	RU	Roughness	5
Corrugations	DVI/UR 2	CO	Roughness/Drainage	5
Camber/crossfall	DVI/UR 3	CR	Drainage	5
Gravel thickness	DVI/UR 4	GR	Structure	5
Erosion gullies	DVI/UR 5	EG	Structure	5
Potholes	DVI/UR 6	PO	Structure	5
Clay	DVI/UR 7	CY	Safety	3

Note: In the case of the example in Figures A.10 and A.11, no unpaved roads are shown.

Figure A.9. **SUMMARY OF ROAD CONDITION SURVEY RESULTS: AVERAGE CONDITION VALUE ("FORM IV")**

ROAD CONDITION SURVEY (RCS)

PAVED AND UNPAVED ROADS

OFFICE: R.D. MAINT. | DISTRICT: MOROKULI

Established by: MUTALI

Date: 88 01 21

PAGE 1

DESIGNATION OF ROAD SECTION									SURVEY DATA			
Road		Section No.	Identification Points (IP)			Section length	Pavement width	AADT	Avg. condition value		Road signs & Furniture	Notes / Remarks
Class	No.		I.P.	Designation	Km	Km	m	Veh/day	Carriageway	Road side elements		
1	2	3	4	5	6	7	8	9	11	12	13	14
D	201	1	From V	080 (Ennoryep)	From 223.50	7.05	6.50	300	3.125	1.9		DVI required
			To J	022	To 216.45							
			From		From							
			To		To							
			From		From							
			To		To							
			From		From							
			To		To							
			From		From							
			To		To							
			From		From							
			To		To							
			From		From							
			To		To							
			From		From							
			To		To							
			From		From							
			To		To							
			From		From							
			To		To							

A.3.2. Parameters for DVI of paved roads, Form VII

Type	No.	Code	Function affected	Points in scale
Rutting without cracks	DVI/PR 1a	RU	Structure/Roughness/ Drainage	5
Rutting with cracks	DVI/PR 1b	RU/C	Structure/Roughness/ Drainage	5
Corrugations	DVI/PR 2	CO	Carriageway	5
Depressions	DVI/PR 3	DP	Drainage	5
Transverse cracks	DVI/PR 4a	CT	Structure	5
Longitudinal cracks	DVI/PR 4b	CL	Structure	5
Alligator cracks	DVI/PR 4c	CA	Structure	5
Holes	DVI/PR 5	HO	Structure	5
Edge distress	DVI/PR 6	ED	Structure	5
Stripping/fretting/ ravelling	DVI/PR 7a	ST	Structure	5
Stripping/fretting/ ravelling of surface	DVI/PR 7b	ST/S	Structure	5
Bleeding	DVI/PR 8	BL	Safety	3

A.3.3. Inspection procedure

The detailed visual inspection should be performed by the same team that performs the road condition survey.

The inspection is carried out by walking the complete section. One of the inspectors inspects the right hand side of the road the other inspects the left hand side. One of them completes the form.

During the inspection the distance will be measured from the starting point onwards.

Sub-sections of earlier determined length will be inspected separately.

The values to be marked on the forms should be discussed and agreed upon by both inspectors. The value representing the severity and extent of the deficiency is stated in the Damage Sheet (Annex C).

A.3.4. Summary of results

The results of the inspection for paved roads are given in Figure A.10 and Figure A.11 which includes a summary of the average condition values.

Figure A.10. **DETAILED VISUAL INSPECTION FORM FOR PAVED ROADS ("FORM VII")**

DETAILED VISUAL INSPECTION (DVI)			**PAVED ROADS**
OFFICE: R.D. MAINT.	DISTRICT: MOROKULI	ROAD DENOMINATION / CLASS: D201	Main Section No. 1
Name of inspector:		FROM I.P.: 080 (Ennoryep)	Km: 223.50
Date: 88 01 23	Weather: ☒ Clear o Rainy; Carr.way: ☒ Dry o Drying o Wet	TO I.P.: 022	Km: 216.45
		Pavement type: DST; Pavement width: 6.50	Section length: 7.05Km

SUB-SECTION		1						2						3						4						5					
Roadside		L			R			L			R			L			R			L			R			L			R		
Damage type - Cat	E \ S	L	M	S	L	M	S	L	M	S	L	M	S	L	M	S	L	M	S	L	M	S	L	M	S	L	M	S	L	M	S
Rutting - PR1	<10%	1	(3)	5	1	3	5	1	(3)	5	1	3	5	1	(3)	5	1	(3)	5	1	3	5	1	3	5	1	3	5	1	3	5
	10-50%	2	4	5	(2)	4	5	2	4	5	2	4	5	2	4	5	2	4	5	2	4	5	(2)	4	5	2	4	5	2	4	5
	>50%	3	5	5	3	5	5	3	5	5	(3)	5	5	3	5	5	3	5	5	(3)	5	5	3	5	5	3	5	5	3	5	5
Corrugations - PR2	<10%	(1)	3	4	(1)	3	4	(1)	3	4	(1)	3	4	1	3	4	1	3	4	1	3	4	1	3	4	1	3	4	1	3	4
	10-50%	2	3	5	2	3	5	2	3	5	2	3	5	(2)	3	5	(2)	3	5	2	3	5	2	3	5	2	3	5	2	3	5
	>50%	3	4	5	3	4	5	3	4	5	3	4	5	3	4	5	3	4	5	(3)	4	5	(3)	4	5	3	4	5	3	4	5
Depressions - PR3	<10%	1	3	4	1	(3)	4	1	3	(4)	1	(3)	4	1	3	4	1	(3)	4	1	3	4	1	3	4	1	3	4	1	3	4
	10-50%	2	(4)	5	2	4	5	2	4	5	2	4	5	2	(4)	5	2	4	5	2	4	5	2	4	5	2	4	5	2	4	5
	>50%	3	5	5	3	5	5	3	5	5	3	5	5	3	5	5	3	5	5	3	(5)	5	(3)	5	5	3	5	5	3	5	5
Transverse cracking - PR4a	<2 (No./100m)	1	3	4	1	3	(4)	1	3	4	1	3	4	1	3	4	1	3	4	1	3	4	1	3	4	1	3	4	1	3	4
	2-15	2	3	(5)	2	3	5	2	3	5	2	3	5	2	3	5	2	3	5	2	3	5	2	3	(5)	2	3	5	2	3	5
	>15	3	4	5	3	4	5	3	(4)	5	(3)	4	5	3	(4)	5	3	4	(5)	3	(4)	5	3	4	5	3	4	5	3	4	5
Longitudinal cracking - PR4b	<10%	1	3	5	1	3	5	1	3	5	1	3	5	1	3	5	1	3	5	1	3	5	1	3	5	1	3	5	1	3	5
	10-50%	2	4	5	2	4	5	2	4	5	2	4	5	2	4	5	2	4	5	2	4	5	2	4	5	2	4	5	2	4	5
	>50%	3	5	5	3	5	5	3	5	5	3	5	5	3	5	5	3	5	5	3	(5)	5	3	5	5	3	5	5	3	5	5
Alligator cracking - PR4c	<10%	1	4	5	1	4	5	1	4	5	1	4	5	1	4	5	1	(4)	5	1	4	5	1	(4)	5	1	4	5	1	4	5
	10-50%	2	5	5	(2)	5	5	2	5	5	2	5	5	2	(5)	5	2	5	5	2	(5)	5	2	5	5	2	5	5	2	5	5
	>50%	(3)	5	5	3	5	5	3	(5)	5	(3)	5	5	3	5	5	3	5	5	3	5	5	3	5	5	3	5	5	3	5	5
Holes - PR5	<5 (No./100m)	1	3	5	1	3	5	1	3	5	1	3	5	1	3	5	1	3	5	1	3	5	1	(3)	5	1	3	5	1	3	5
	5-15	3	4	5	3	4	5	3	4	5	3	4	5	3	4	5	(3)	4	5	3	(4)	5	3	4	5	3	4	5	3	4	5
	> 15	(4)	5	5	4	(5)	5	(4)	5	5	4	(5)	5	(4)	5	5	4	5	5	4	5	5	4	5	5	4	5	5	4	5	5
Edge distress - PR6	<10%	1	(3)	4	1	3	4	1	3	4	1	3	4	1	3	4	1	3	4	1	3	4	1	3	4	1	3	4	1	3	4
	10-50%	2	3	5	(2)	3	5	2	3	5	(2)	3	5	2	3	5	(2)	3	5	2	3	5	2	3	5	2	3	5	2	3	5
	>50%	3	4	5	3	4	5	(3)	4	5	3	4	5	(3)	4	5	3	4	5	3	(4)	5	(3)	4	5	3	4	5	3	4	5
Stripping/ Fretting/ Ravelling - PR7a	<10%	1	2	4	1	2	4	1	(2)	4	1	(2)	4	1	(2)	4	1	(2)	4	1	2	4	1	2	4	1	2	4	1	2	4
	10-50%	1	3	5	1	3	5	1	3	5	1	3	5	1	3	5	1	3	5	1	3	5	1	3	5	1	3	5	1	3	5
	>50%	(2)	5	5	(2)	5	5	2	5	5	2	5	5	2	5	5	2	5	5	2	5	5	2	5	5	2	5	5	2	5	5
Stripping/Fretting/ Ravelling of surface - PR7b	<15%	1	2	3	1	2	3	1	2	3	1	2	3	1	2	3	1	2	3	1	2	3	1	2	3	1	2	3	1	2	3
	15-30%	2	4	5	2	4	5	2	4	5	2	4	5	2	4	5	2	4	5	2	4	5	2	4	5	2	4	5	2	4	5
	>30%	3	5	5	3	5	5	3	5	5	3	5	5	3	5	5	3	5	5	3	5	5	3	5	5	3	5	5	3	5	5
Bleeding - PR8	< 5%	(1)			(1)			1			(1)			(1)			(1)			1			1			1			1		
	5-50%	2			2			(2)			2			2			2			2			2			2			2		
	>50%	3			3			3			3			3			3			(3)			(3)			3			3		

Remarks: ____________________

Figure A.11. **SUMMARY OF DETAILED VISUAL INSPECTION RESULTS: AVERAGE CONDITION VALUE "FORM VIII"**

DETAILED VISUAL INSPECTION (DVI)

OFFICE: R.D. MAINT.	DISTRICT: MOROKULI	PAGE 1
Established by: MUTALI		
Date: 88 01 24		

DESIGNATION OF ROAD SECTION

Road Class	Road No.	Section / Sub-section No.	Identification Points (IP) I.P.	Designation	Km	Section length Km	Carriageway width m	AADT Veh/day
1	2	3	4	5	6	7	8	9
D	201	1.1	From V	080 (Ennoryep)	223.50	1.50	6.50	300
			To CV		222.00			
		1.2	From CV		222.00	3.45	6.50	300
			To CC		218.55			
		1.3	From CC		218.55	1.20	6.50	300
			To CP		217.35			
		1.4	From CP		217.35	0.90	6.50	300
			To J	022	216.45			
			From					
			To					
			From					
			To					
			From					
			To					
			From					
			To					
			From					
			To					
			From					
			To					

Roadside	Unpaved roads: Rutting	Corrugations	Loss of camber	Gravel thickness	Erosion gullies	Potholes	Clay		Paved roads: Rutting	Corrugations	Depressions	Tr/Long cracking	Alligator crack.	Holes	Edge distress	Stripping	Bleeding	Avg.cond.value	Instrumental survey data: IRI	Deflection	Skid resistance
11	12	13	14	15	16	17	18	19	20	21	22	23	24	25	26	27	28	29	30	31	32
L									3	1	4	5	3	4	3	2	1	2.7			
R									2	1	3	4	2	5	2	2	1				
L									3	1	4	4	5	4	3	2	2	2.8			
R									3	1	3	3	3	5	2	2	1				
L									3	2	4	4	5	4	3	2	1	2.9			
R									3	2	3	5	4	3	2	2	1				
L									3	3	5	4	5	4	4	4	3	3.5			
R									2	3	3	5	4	3	3	2	3				
L																					
R																					
L																					
R																					
L																					
R																					
L																					
R																					
L																					
R																					
L																					
R																					

Annex B

ASSESSMENT OF ROAD SURFACE ROUGHNESS (EVENNESS)

B.1. INTERNATIONAL ROUGHNESS INDEX: IRI

Riding quality is without doubt a most important factor for the road user, influencing both user comfort and also vehicle operating costs. Roughness, together with surface distress, is a determining parameter affecting the decision on additional visual inspections at specific parts of the network.

The international roughness index (IRI) is a reference measure of the road surface shape that resulted from an international co-operative study, through the World Bank. The IRI is a measure of the impact of the road profile on the ride and dynamic response of a moving vehicle. It is defined by a mathematical processing of the absolute longitudinal profile of road surface elevations in a wheelpath, simulating the motion of a wheel travelling at 80 km/h. The IRI is expressed in units of m/km, representing average displacements per unit distance. The IRI can be determined in different ways, as follows:

-- Directly from measures of the longitudinal profile by a standard mathematical processing of elevation data; or

-- Indirectly, by measuring the displacement response of an instrumented roadmeter (RTRRMS) calibrated to IRI; or

-- Indirectly, by subjective estimation in a way which is considered to provide sufficient accuracy for distinguishing five different classes for the road condition survey.

Annex D deals with the first two ways. The third, a quick approximate method, is described in the following section.

B.2. SUBJECTIVE DETERMINATION OF THE IRI

The method presented for the determination of the IRI is derived from World Bank recommendations for cases where more refined methods are not appropriate. This method is applicable where an approximation is sufficient which is the case in this manual.

The method enables the estimation of the roughness in IRI units by comparing different subjective indicators with the corresponding scale of the IRI. Tables B.1. and B.2. show the relationships for unpaved and paved roads respectively. The subjective estimators used to estimate roughness are:

-- Typical speed;
-- Driving condition;
-- Surface damage.

Typical speed

This is the normal design speed on a dry and straight road without interference by the other road users and adapted to the road condition in such a way as to safeguard the mechanical integrity of the vehicle as well as the safety and comfort of the passengers of the vehicle.

Driving conditions

In Tables B.1 and B.2 the driving sensations (comfort) are related to a particular medium-sized vehicle with an independent suspension system. Given the differences which may exist between vehicles, the inspector is to derive his own calibration based on his own vehicle. The driving sensations can be described in simple expressions in relation to abrupt vibrations or movements observed by the inspector when driving at a given speed over the road.

This description should be adapted to local conditions and vehicle type.

Surface damage

The surface defects included in each description are typical examples of how the surface condition may appear at each roughness level, for surfaces which were originally of good shape. Note that only defects which distort the shape of the surface are taken into account, and other types of visible distress (such as cracking) are ignored for the purposes of estimating the roughness. The severity of the shape defects is based on depth and on the occurrence, defined by three levels, namely: occasional; moderate; frequent.

Adaptation of the inspection method

To enable the application of this method for road inspection, the scale given in Tables B.1 and B.2 is divided into five categories (related to the scale adopted in this document) representing the five road conditions recorded:

Condition rating	Maintenance activity required "unpaved roads"
1 = excellent; no damage visible	No action required.
2 = good	No immediate action required.
3 = fair	Dragging and/or grading; in some cases the processing of the base course may be required.
4 = critical	Processing of existing base course and/or adding of gravel/selected material necessary.
5 = failed	Adding gravel or selected material and processing of base course.

Note: In general a rating of 3 or higher implies that a Detailed Visual Inspection is required to reveal the proper type of maintenance activity required.

Table B.1

IRI Classification (m/km)	UNPAVED ROADS	OECD Classification
0 2 4	Recently bladed surface of fine gravel, or soil with excellent longitudinal and transverse profile (usually only found in short lengths). Ride comfortable up to 80-100 km/h, aware of gentle undulations or swaying. Negligible depressions (e.g. < 5 mm/3 m) and no potholes.	1
6 8 10	Ride comfortable up to 70-80 km/h but aware of sharp movements and some wheel bounce. Frequent shallow - moderate depressions or shallow potholes (e.g. 6-30 mm/3 m with frequency of 5-10 per 50 m). Moderate corrugations (e.g. 6-20 mm/0.7-1.5 m).	2
12 14	Ride comfortable at 50 km/h, or on specific sections 40-70 km/h. Frequent moderate transverse depressions (e.g. 20-40 mm/3-5 m at frequency 10-20 per 50 m) or occasional deep depressions or potholes (e.g. 40-80 mm/3 m with frequency less than 5 per 50 m). Strong corrugation (e.g. > 20 mm/0.7-1.5 m).	3
16 18	Ride comfortable at 30-40 km/h. Frequent deep transverse depressions and/or potholes (e.g. 40-80 mm/3-5 m at frequency 5-10 per 50 m); or occasional very deep depressions (e.g. 80 mm/1-5 m with frequency less than 5 per 50 m) with other shallow depressions. Not possible to avoid all the depressions but the worst.	4
20 22 24	Ride comfortable at 20-30 km/h. Speeds higher than 40-50 km/h would cause extreme discomfort and possible damage to a car. On a good general profile; frequent depressions and/or potholes (e.g. 40-80 mm/1-5 m at a frequency of 10-15 per 50 m) and occasional very deep depressions (e.g. 80 mm/0.6-2 m). On a poor general profile; frequent moderate defects and depressions (e.g. poor earth surface).	5

Table B.2

IRI Classification (m/km)	PAVED ROADS	OECD Classification
0 2	V* > 120 km/h Undulations hardly noticeable at 80 km/h. Visual inspection: No deformations or potholes. Depressions < 2 mm/3 m Asphalt concrete or high quality surface treatment.	1
4 5 6	V = 100-120 km/h. Aware of slight movements or gentle undulations at 80 km/h. Surface degraded by: depressions (5-15 mm/3 m), repairs and potholes (1-5 per 100 m). Surface not damaged, but undulations and corrugations exist.	2
7	V = 70-90 km/h Sharp movements and swaying. Depressions (10-20 mm) or frequent repairs, or occasional potholes; or surface not damaged, but strong undulations and corrugations exist.	3
9 10	V = 50-60 km/h Frequent and sudden movements and swaying. Serious faults; or frequent and deep depressions (20-35 mm at a frequency of 6-10 per 100 m); or frequent poor repairs and potholes (6-20 per 100 m)	4
11	V < 50 km/h. Frequent and deep depressions and potholes (> 40 mm; frequency 16-30 per 100 m).	5

*V = "comfortable" speed.

Examining this figure shows that the surface can be monitored in five categories, all related to typical speeds:

	Unpaved Roads	Paved Roads
-- 1	= 80-100 km/h	> 120 km/h
-- 2	= 70-80 km/h	100-120 km/h
-- 3	= 40-70 km/h	70-90 km/h
-- 4	= 30-40 km/h	50-60 km/h
-- 5	= < 30 km/h	< 50 km/h

On the other hand it is possible to distinguish five comfort categories:

-- 1 = excellent
-- 2 = good
-- 3 = fair
-- 4 = poor
-- 5 = none.

Taking into account factors such as topography/longitudinal profile, horizontal alignment (curves, etc), and interference of other road users, one may use a formula, given the relationship between the speed and the comfort, in which the importance of the comfort factor has a greater influence than the speed factor.

For example, one may use the following formula:

$U = (2*C + V)/3$, in which
U = OECD classification
C = Comfort parameter ($1 < C < 5$)
V = Speed parameter ($1 < V < 5$)

This formula gives the following conversion matrix to obtain U from C and V:

Comfort Category 'C'	Speed 'V' (km/h)					
	Paved	> 120	100-120	70-90	50-60	< 50
	Unpaved	80-100	70-80	40-70	30-40	20-30
		1	2	3	4	5
1		1	1	2	2	3
2		2	2	2	3	4
3		2	3	3	3	4
4		3	3	4	4	5
5		3	4	4	5	5

The condition rating on the diagonal (top left to bottom right) corresponds with the rating given in Table B.1 (straight and dry road, no interference).

The elements above the diagonal correspond to the case where the speed is limited by factors not related to the road surface condition.

The elements below the diagonal are only encountered in exceptional cases (except when potholes are present or alligator cracks have appeared) since they represent situations where the speed is not adapted to the road surface condition. It should be noted that corrugations tend to increase speeds, whereas potholes tend to reduce the speed.

Annex C

SUMMARY OF DAMAGE CATALOGUE

C.1. INTRODUCTION

To obtain evaluations as objective as possible and which are suitable for qualified personnel, it is mandatory to base visual inspections/surveys on a reference document such as a damage catalogue. The catalogue presented in a separate booklet comprises a number of sheets corresponding to those parameters which are evaluated and damage which most frequently occur in paved and unpaved roads. Figure C.1 shows the position of the damage catalogue in the general scheme.

The proposed survey/inspection procedure includes two evaluation levels requiring a precise identification and quantification of principal defects. Although high output measuring equipment, such as the "Bump Integrator" or the "Longitudinal Profile Analyser" (APL) (see Annex D), may be used for a quick inspection of the road system, this manual is based on visual surveys/ inspections, completed by simple measurements.

The damage is divided into that inspected during the Road Condition Survey (RCS) and that inspected during the Detailed Visual Inspection (DVI).

C.2. DAMAGE SHEETS

For each type of damage, the catalogue contains a sheet consisting of two parts:

A. The first part contains information relative to surveys/inspections:

- -- A photo or illustration if possible of the damage in various stages;
- -- A detailed description of the damage;
- -- A description of inspection and/or proposed measurement methods;
- -- Quantitative criteria to record the extent and/or severity of the damage.

B. The second part contains the following information:

- -- Probable cause of the damage, preliminary symptoms and consequences, if no corrective maintenance actions are carried out;

Figure C.1. **POSITION OF THE DAMAGE CATALOGUE IN THE GENERAL SCHEME**

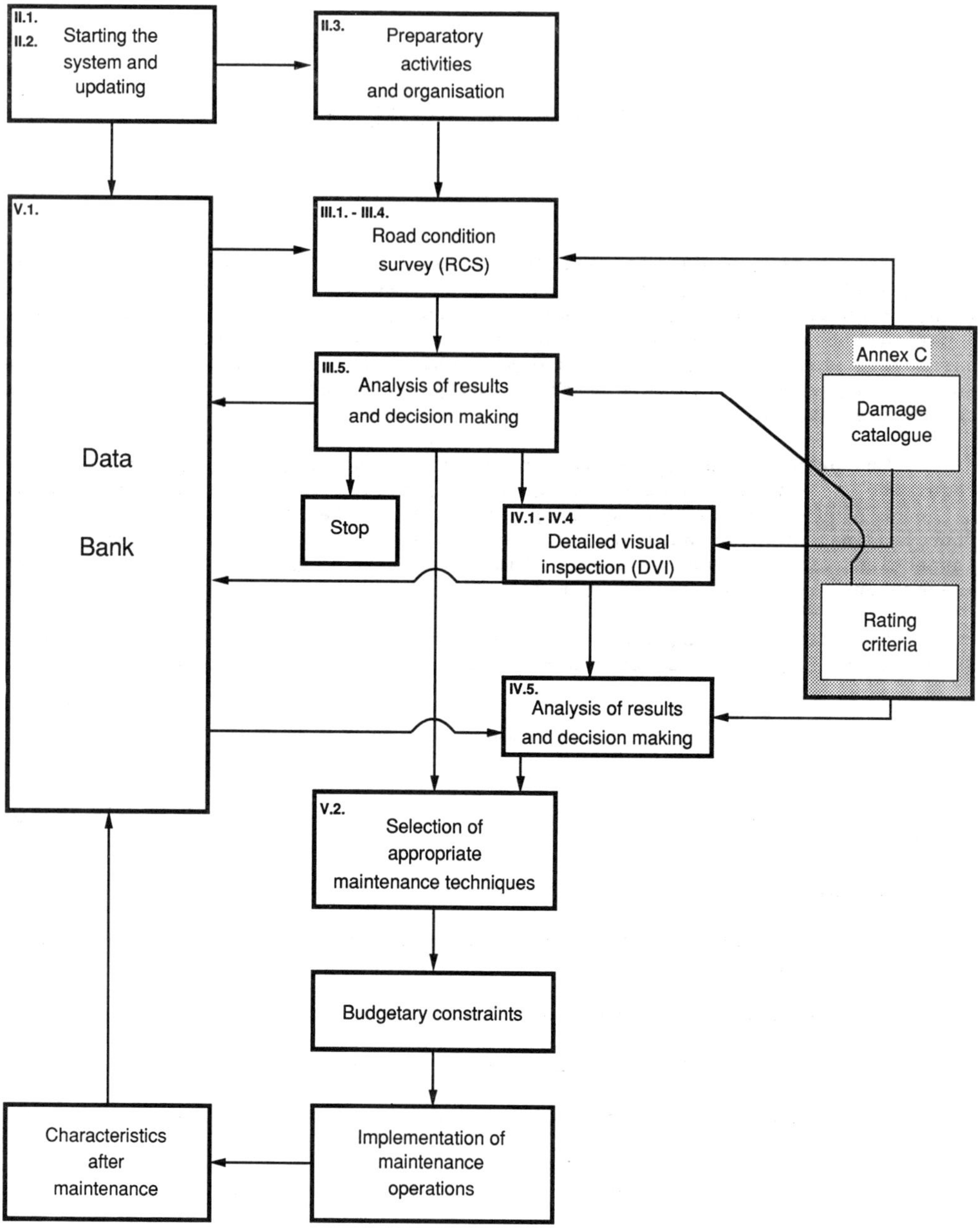

-- Factors affecting the development of the damage such as climate, traffic and materials;

-- Proposed operations or remedies (divided into specific tasks), and taking into account the various causes leading to the present situation.

The last section of the sheet is indicative only and the information given is summarised and may be incomplete. As such it should not be used by the maintenance department to define the maintenance activity to be undertaken. There are other publications that deal more specifically with maintenance and rehabilitation techniques and it is recommended to use these.

C.3. RATING OF DAMAGE

C.3.1. Road Condition Survey (RCS)

The classification of the carriageway or road surface condition is done on a five-point scale, based upon the criteria given in Annex B.

In addition to the damage affecting the road surfacing, the inspectors also evaluate the defects and damage of the drainage system and the area adjacent to the road (fill, etc.). This is done with a three-point rating system for roadside elements, road signs and furniture and structures in accordance with the description given on the relevant damage sheet.

The principles of this three-point system are that the values recorded for each element reflect the efficiency of routine and recurrent maintenance activities. The need for immediate intervention is based on considerations relating to road user safety and the stability of the road, embankment and structures. It is recorded in the "Remarks" section.

RATING CRITERIA

Value	Routine maintenance	Action
1	Satisfactory	Not necessary
2	To be checked	To be expected shortly
3	Not satisfactory	Necessary

Table C.1

ROAD CONDITION SURVEY: ROAD PARAMETERS

Group	Type	No. of sheet	Rating levels	Section*
Carriageway	Surface quality	-	5	Annex B
	Prevailing damage (if necessary)	-	-	3-4
Shoulders	Deformation	RCS 1	3	1
	Scour	RCS 2	3	
Side drains	Siltation	RCS 3	3	1
	Scour	RCS 4	3	
Road	Debris/vegetation encroachment	RCS 5	3	1
	Obstacles	RCS 6	3	
Road signs & furniture; Road markings	Damaged and/or dirty, missing	RCS 7	3	1
Damage & defects to be repaired immediately	Various, to be mentioned under "Remarks"	-	-	3-4

Table C.2

ROAD CONDITION SURVEY: STRUCTURES

Group	Type	No. of sheet	Rating levels	Section*
Culverts	Silted/blocked	RCS 8	3	2
	Scour	RCS 9	3	
	Structural damage	RCS 10	3	
Bridges	Silted/blocked	RCS 11	3	2
	Scour	RCS 12	3	
	Structural damage	RCS 13	3	

* Numbers refer to Sections of the Damage Catalogue, published separately.

C.3.2. Detailed Visual Inspection (DVI)

The rating of damage affecting the carriageway is based on two characteristics: extent and severity.

The extent may be defined as that part of the road affected by the damage.

For linear damage (rutting, edge distress, longitudinal cracks) the inspector estimates the length of the damage in proportion to the length of the sub-section.

For two-dimensional damage (corrugations, alligator cracking, peeling, bleeding, etc.) the inspector estimates the damaged area in proportion to the total area of the sub-section.

For both cases the proposed method distinguishes only three levels. These levels differ for the various types of defects and are given on the relevant sheet.

In this method, the severity of the damage is expressed as:

1. light;
2. moderate;
3. severe.

Purely superficial damage such as deposits of clay (CY) on unpaved roads or bleeding (BL) on paved roads cannot be quantified in terms of severity.

The significance of the damage is derived from a combination of the extent and severity. This is rated on a five-point scale established in matrices presented on the damage sheets. Apart from the defects CY and BL (assessed on the basis of their extent), all defects are assigned a value in accordance with this five-point rating scale.

Note:

-- It is of the utmost importance that the inspection teams have an intimate knowledge of the information contained in the damage catalogue.

-- It is strongly recommended that, prior to actual inspection work, teams receive proper training related to the inspection programme (see Chapter VI).

Table C.3

DETAILED VISUAL INSPECTION OF UNPAVED ROADS

Type	No. of sheet	Code	Function affected	Rating levels
Rutting	DVI/UR 1	RU	Roughness/ Drainage	5
Corrugations	DVI/UR 2	CO	Roughness	5
Camber/crossfall	DVI/UR 3	CR	Drainage	5
Gravel thickness	DVI/UR 4	GR	Structure	5
Erosion gullies	DVI/UR 5	EG	Structure	5
Potholes	DVI/UR 6	PO	Structure	5
Clay	DVI/UR 7	CY	Safety	3

Table C.4

DETAILED VISUAL INSPECTION OF PAVED ROADS

Type	No. of sheet	Code	Function affected	Rating levels
Rutting without cracks	DVI/PR 1a	RU	Structure/ Roughness/ Drainage	5
Rutting with cracks	DVI/PR 1b	RU/C	Structure/ Roughness/ Drainage	5
Corrugations	DVI/PR 2	CO	Carriageway	5
Depressions	DVI/PR 3	DP	Drainage	5
Transverse cracks	DVI/PR 4a	CT	Structure	5
Longitudinal cracks	DVI/PR 4b	CL	Structure	5
Alligator cracks	DVI/PR 4c	CA	Structure	5
Holes	DVI/PR 5	HO	Structure	5
Edge distress	DVI/PR 6	ED	Structure	5
Stripping/fretting/ ravelling	DVI/PR 7a	ST	Structure	5
Stripping/fretting/ ravelling of surface	DVI/PR 7b	ST/S	Structure	5
Bleeding	DVI/PR 8	BL	Safety	3

Annex D

OVERVIEW OF SPECIAL MEASURING EQUIPMENT

D.1. RATIONALISING AND ENHANCING VISUAL INSPECTIONS

This manual and its proposed monitoring practice are basically built on visual inspections involving simple tools, as indicated in Section II.3.5. It was the very aim of this project to develop a monitoring scheme that is simple, straightforward and easy to apply, without employing sophisticated equipment. The methods proposed can, however, be relatively time consuming and labour intensive.

A wide range of high output measuring instruments is now available to assist the engineer. These require substantial financial resources, both for their purchase and their maintenance, as well as highly qualified personnel.

Some of this equipment operates fully automatically, and thus improves the efficiency of the inspection operation. However, the requirements and work entailed in processing and analysing the measured data should not be underestimated.

The high efficiency measuring instruments can be used to complement the Detailed Visual Inspection method -- presented in the Manual --, thus reducing possible subjectivity, incompleteness or inaccuracy. It is evident that the results of these measurements need to be fitted into the five-point evaluation rule of this Manual. Additional tests and/or assessments can easily be introduced into the inspection system and be treated together with the data obtained through the basic monitoring process.

In the following, selected high output measuring equipment is presented with emphasis on roughness ("evenness"), structural performance ("deflection") and friction properties ("skid resistance"). The overview given hereunder is not exhaustive, but it may give some indication on the type of equipment that has been and is used in practice (not only in research), together with approximate cost estimates.

For further reference, the reader is advised to consult:

-- The OECD reports on:

- Pavement management systems (1987);
- Road surface characteristics (1984);
- Maintenance techniques for road surfacings (1978);
- Road strengthening (1976); and
- Maintenance of rural roads (1973).

(available from: OECD, Road Transport Research
2, rue André-Pascal
75775 Paris Cedex 16)

-- The PIARC Technical Committees' reports on this subject prepared for the World Congresses in Brussels, 1987; Sydney, 1983; Vienna, 1979; and Mexico City, 1975.

(available from: AIPCR/PIARC
27, rue Guénégaud
F-75006 Paris)

Also, the reader may contact the institutions and Members listed at the end of this report.

D.2. HIGH OUTPUT EQUIPMENT FOR MEASURING ROUGHNESS ("EVENNESS")

The International Roughness Index (IRI) can be determined from the following methods:

a) The inspection method presented in Annex B;

b) Measurements of the geometric profile of the road; and

c) Measurements with dynamic response instruments.

There are apparatus using a geometric reference, such as the Viagraph and the AASHO Slope Profilometer (low speed), the British High Speed Profilometer, etc. High efficiency instruments based on the inertia principle include, the French Analyseur de Profil en Long (Longitudinal Profile Analyser -- APL 72 and, more recently, APL Unique) and the GM Profilometer (for research purposes). As an illustration of the financial resources required, the British High-Speed Road Monitor -- a high speed profilometer measuring longitudinal profile, macrotexture (see D.4) and other parameters -- costs approximately US$ 375 000 for purchase and US$ 2 300 to 3 000 per day for survey service. The APL trailer with its data processing unit costs approximately US$ 50 000 (VAT excluded); its performance is 150 km/day at a cost of US$ 30 per kilometre inspected.

Apparatus measuring the vertical displacements of a moving vehicle include the well-known British Bump Integrator, the American Mays Ride Meter, and the Australian NAASRA Meter. Equipment measuring vertical acceleration includes the Swedish Road Surface Tester, which measures the acceleration of an axle or a sensor carrying arm. Others reconstitute the movements of the chassis or the vehicle axle, including the Canadian Automatic Road Analyser (ARAN). Approximate costs for a Bump Integrator are US$ 37 500 for the purchase of the trailer and the microprocessor unit, and US$ 1 150 per day for survey service excluding processing costs. The ARAN, which measures various profile curvature and other parameters, has a purchase cost of US$ 250 000 to 400 000, depending on outfit; its performance is 250 km/day.

D.3. HIGH OUTPUT EQUIPMENT FOR MEASURING STRUCTURAL PERFORMANCE

D.3.1. Structural integrity: deflections

There are three main types of equipment used in assessing structural performance based on deflection measurements, where deflection is the vertical displacement of a point of the surface of a pavement when a load passes over it.

A distinction is made between static deflection [see a) below] and dynamic deflection [see b) and c) below]. Each of the equipment measures the "elastic" and/or "rebound" response at the pavement surface when it is subjected to a known load. It is assumed that the magnitude of the response can determine the overall condition and the remaining life of the pavement.

Representative pieces of the three equipment types are :

a) The well-known, fully automated Deflectograph La Croix, initially developed in France and now used in many countries, which provides high output Benkelman Beam measurements. Approximate costs for a Deflectograph La Croix are, US$ 200 000 to 250 000 for purchase, and US$ 180 per kilometre for survey service, with a performance of 12-18 km per day;

b) The Falling Weight Deflectometer (FWD) measures the vertical deflection of a pavement under an impact load. Approximate costs for a Falling Weight Deflectometer which is now used in several European countries and in the United States are, US$ 75 000 to 150 000 for purchase, depending on outfit, and US$ 1 500 to 2 300 per day for survey service.

c) The Dynaflect and the Road Rater are lightweight vibrators for routine use, mainly applied in the United States. They have been developed for rapid deflection measurements under fairly small dynamic loads at given frequencies, and their results are correlated with the deflection measured by the Benkelman Beam. Approximate costs of a Dynaflect are around US$ 25 000 and US$ 15 in operating costs per lane kilometre surveyed

D.3.2. Individual layers

It should be realised that deflection measurements evaluate the whole structure as one unit. However, the failure of a pavement may be due to the poor performance of one of its constituent layers. To determine the real cause, the conditions of individual layers must be assessed by the help of boreholes, pits or trenches, and by appropriate testing of materials and strengths, in addition to determining real layer thicknesses. To overcome the inconvenience entailed in these operations, the Dynamic Cone Penetrometer (DCP) has been developed which may give some useful indications about the status of the lower layers.

D.4. EQUIPMENT FOR MEASURING SKID RESISTANCE

Macrotexture (or macroroughness), i.e. the surface relief of the pavement, is best determined through the Sand Patch method. Microtexture (or microroughness), indicating the degree of polishing of a pavement surface, is evaluated most frequently from the SRT -- Skid Resistance Tester (Pendulum). These tests do obviously not provide high output results, but are mentioned here because of their importance in practice.

The main high output equipment for measuring skid resistance are, the British SCRIM (approx. cost: US$ 150 000), the United Stated ASTM Skid-tester (US$ 50 000 to 100 000, depending on specifications), the Canadian ASTM Skid-trailer, the German Stuttgarter Reibungsmesser, the French LPC-Remorque de glissance, the Dutch RED-Trailer, the Japanese Skid Resistance Measuring Machine, and the Swiss Skiddometer BV-11. Survey service costs per kilometre of measurements undertaken by SCRIM are between US$ 15 and 30 depending on conditions of use.

D.5. FINAL REMARK

Since the maintenance and rehabilitation of road networks is becoming an increasingly serious problem worldwide, economic and efficient inspection techniques and management strategies are needed in order to ensure optimum road surface conditions and all-weather serviceability.

The OECD Road Transport Research Programme is therefore launching a new project on maintenance management systems in developing countries, in the framework of which more detailed proposals for applications of high output measuring equipment will be given.

LIST OF GROUP MEMBERS

Chairman: Mr. Geir REFSDAL

Australia	Dr. J. METCALF Australian Road Research Board (ARRB) P.O. Box 156 (Bag 4) Nunawading, Victoria 3131
Belgium	Dr. Louis FRANCKEN Chercheur au service "Conception et structures" Centre de Recherches Routières Boulevard de la Woluwe 42 B-1200 Brussels
	Dr. Michel GORSKI Chercheur - Service Etudes Générales Centre de Recherches Routières Boulevard de la Woluwe 42 B-1200 Brussels
	Mr. Jean LECLERCQ Chargé de cours à la Faculté des Sciences agronomiques de l'Etat Service "Résistance des matériaux et construction du génie rural" B-5800 Gembloux
Denmark	Mr. Jan JANSEN National Road Laboratory Elisagaardsvej 5 DK-4000 Roskilde
Finland	Mr. Lars BJÖRKSTEN Roads and Waterways Administration (RWA) PL 33 SF-00521 Helsinki
France	Mr. C. PEYRONNE SETRA 46 Avenue Aristide Briand 92223 Bagneux

	Mr. G. CAROFF SCETAUROUTE Technique et Développement 2 rue Stephenson 78181 St. Quentin-Yvelines Cédex
<u>Germany</u>	Mr. H. NEUBAUER Baudirektor Kreditanstalt für Wiederaufbau Hauptabteilung Technik Palmengarlenstr. 5-9 D-6000 Frankfurt am Main
<u>Ireland</u>	Mr. J. SHEEDY Head, Road Construction Section An Foras Forbartha Pottery Road Deansgrange Co. Dublin
<u>Italy</u>	Ing. E. SCOTTO Chief Engineer ANAS Via Monzambano 10 I-00185 Rome
<u>Japan</u>	Mr. Tomonobu NAKAOKA Deputy Director International Affairs Division Ministry of Construction 2-1-3 Kasumigaseki, Chiyoda-ku Tokyo 305
<u>Netherlands</u>	Ing. H. VAN DEN BELD Senior Engineer DHV Consulting Engineers Laan 1914 Nr. 35/P.O. Box 85 3800 AB Amersfoort
	Mr. A. BRESSER M.Sc. DHV Consulting Engineers Project Director P.O. Box 85 3800 AB Amersfoort
<u>Norway</u>	Mr. Geir REFSDAL Senior Engineer Norwegian Directorate of Roads Road Research Laboratory P.O. Box 6390 Etterstad N-0604 Oslo 6

<u>Turkey</u>	Mr. Mehmet A.ÖNAL Pavement Design Eng. Karayollari Genel Müdürlügü Üstyapi Subesi Müdürlügü Ankara
	Mr. A. ISFENDIYAR Ind. Eng. Karayollari Genel Müdürlügü Planlama Subesi Müdürlügü Ankara
<u>United Kingdom</u>	Mr. David BROOKS Head of Pavement Management Section of Overseas Unit TRRL Old Wokingham Road Crowthorne, Berks RG11 6AU
<u>United States</u>	Mr. John CUTRELL Director International Highway Programs (HPI-10) Federal Highway Administration Department of Transportation 400 7th Street, S.W. Washington, D.C. 20590
	Mr. B. LORD Pavement Division (HNR-20) Office of Engineering and Highway Operations Research and Development Federal Highway Administration Department of Transportation 6300 Georgetown Pike McLean, Virginia 22101
<u>World Bank</u>	Dr. W. PATERSON The World Bank 1818 H Street N.W. Washington, D.C. 20433
<u>OECD</u>	Mr. B. HORN Road Transport Research 2, rue André-Pascal 75775 Paris Cedex 16

Distributors of World Bank Publications

ARGENTINA
Carlos Hirsch, SRL
Galeria Guemes
Florida 165, 4th Floor-Ofc. 453/465
1333 Buenos Aires

AUSTRALIA, PAPUA NEW GUINEA, FIJI, SOLOMON ISLANDS, VANUATU, AND WESTERN SAMOA
D.A. Books & Journals
11-13 Station Street
Mitcham 3132
Victoria

AUSTRIA
Gerold and Co.
Graben 31
A-1011 Wien

BAHRAIN
Bahrain Research and Consultancy Associates Ltd.
P.O. Box 22103
Manama Town 317

BANGLADESH
Micro Industries Development Assistance Society (MIDAS)
House 56, Road 7A
Dhanmondi R/Area
Dhaka 1209

Branch office:
156, Nur Ahmed Sarak
Chittagong 4000

BELGIUM
Publications des Nations Unies
Av. du Roi 202
1060 Brussels

BRAZIL
Publicacoes Tecnicas Internacionais Ltda.
Rua Peixoto Gomide, 209
01409 Sao Paulo, SP

CANADA
Le Diffuseur
C.P. 85, 1501B rue Ampère
Boucherville, Quebec
J4B 5E6

CHINA
China Financial & Economic Publishing House
8, Da Fo Si Dong Jie
Beijing

COLOMBIA
Enlace Ltda.
Apartado Aereo 34270
Bogota D.E.

COSTA RICA
Libreria Trejos
Calle 11-13
Av. Fernandez Guell
San Jose

COTE D'IVOIRE
Centre d'Edition et de Diffusion Africaines (CEDA)
04 B.P. 541
Abidjan 04 Plateau

CYPRUS
MEMRB Information Services
P.O. Box 2098
Nicosia

DENMARK
SamfundsLitteratur
Rosenoerns Allé 11
DK-1970 Frederiksberg C

DOMINICAN REPUBLIC
Editora Taller, C. por A.
Restauracion e Isabel la Catolica 309
Apartado Postal 2190
Santo Domingo

EL SALVADOR
Fusades
Avenida Manuel Enrique Araujo #3530
Edificio SISA, ler. Piso
San Salvador

EGYPT, ARAB REPUBLIC OF
Al Ahram
Al Galaa Street
Cairo

The Middle East Observer
8 Chawarbi Street
Cairo

FINLAND
Akateeminen Kirjakauppa
P.O. Box 128
SF-00101
Helsinki 10

FRANCE
World Bank Publications
66, avenue d'Iéna
75116 Paris

GERMANY, FEDERAL REPUBLIC OF
UNO-Verlag
Poppelsdorfer Allee 55
D-5300 Bonn 1

GREECE
KEME
24, Ippodamou Street Platia Plastiras
Athens-11635

GUATEMALA
Librerias Piedra Santa
Centro Cultural Piedra Santa
11 calle 6-50 zona 1
Guatemala City

HONG KONG, MACAO
Asia 2000 Ltd.
6 Fl., 146 Prince Edward Road, W. Kowloon
Hong Kong

HUNGARY
Kultura
P.O. Box 139
1389 Budapest 62

INDIA
Allied Publishers Private Ltd.
751 Mount Road
Madras - 600 002

Branch offices:
15 J.N. Heredia Marg
Ballard Estate
Bombay - 400 038

13/14 Asaf Ali Road
New Delhi - 110 002

17 Chittaranjan Avenue
Calcutta - 700 072

Jayadeva Hostel Building
5th Main Road Gandhinagar
Bangalore - 560 009

3-5-1129 Kachiguda Cross Road
Hyderabad - 500 027

Prarthana Flats, 2nd Floor
Near Thakore Baug, Navrangpura
Ahmedabad - 380 009

Patiala House
16-A Ashok Marg
Lucknow - 226 001

INDONESIA
Pt. Indira Limited
Jl. Sam Ratulangi 37
Jakarta Pusat
P.O. Box 181

IRELAND
TDC Publishers
12 North Frederick Street
Dublin 1

ITALY
Licosa Commissionaria Sansoni SPA
Via Benedetto Fortini, 120/10
Casella Postale 552
50125 Florence

JAPAN
Eastern Book Service
37-3, Hongo 3-Chome, Bunkyo-ku 113
Tokyo

KENYA
Africa Book Service (E.A.) Ltd.
P.O. Box 45245
Nairobi

KOREA, REPUBLIC OF
Pan Korea Book Corporation
P.O. Box 101, Kwangwhamun
Seoul

KUWAIT
MEMRB Information Services
P.O. Box 5465

MALAYSIA
University of Malaya Cooperative Bookshop, Limited
P.O. Box 1127, Jalan Pantai Baru
Kuala Lumpur

MEXICO
INFOTEC
Apartado Postal 22-860
14060 Tlalpan, Mexico D.F.

MOROCCO
Societe d'Etudes Marketing Marocaine
12 rue Mozart, Bd. d'Anfa
Casablanca

NETHERLANDS
InOr-Publikaties b.v.
P.O. Box 14
7240 BA Lochem

NEW ZEALAND
Hills Library and Information Service
Private Bag
New Market
Auckland

NIGERIA
University Press Limited
Three Crowns Building Jericho
Private Mail Bag 5095
Ibadan

NORWAY
Narvesen Information Center
Bertrand Narvesens vei 2
P.O. Box 6125 Etterstad
N-0602 Oslo 6

OMAN
MEMRB Information Services
P.O. Box 1613, Seeb Airport
Muscat

PAKISTAN
Mirza Book Agency
65, Shahrah-e-Quaid-e-Azam
P.O. Box No. 729
Lahore 3

PERU
Editorial Desarrollo SA
Apartado 3824
Lima

PHILIPPINES
National Book Store
701 Rizal Avenue
P.O. Box 1934
Metro Manila

POLAND
ORPAN
Patac Kultury i Nauki
00-901 Warszawa

PORTUGAL
Livraria Portugal
Rua Do Carmo 70-74
1200 Lisbon

SAUDI ARABIA, QATAR
Jarir Book Store
P.O. Box 3196
Riyadh 11471

MEMRB Information Services
Branch offices:
Al Alsa Street
Al Dahna Center
First Floor
P.O. Box 7188
Riyadh

Haji Abdullah Alireza Building
King Khaled Street
P.O. Box 3969
Damman

33, Mohammed Hassan Awad Street
P.O. Box 5978
Jeddah

SINGAPORE, TAIWAN, MYANMAR, BRUNEI
Information Publications Private, Ltd.
02-06 1st Fl., Pei-Fu Industrial Bldg.
24 New Industrial Road
Singapore 1953

SOUTH AFRICA, BOTSWANA
For single titles:
Oxford University Press Southern Africa
P.O. Box 1141
Cape Town 8000

For subscription orders:
International Subscription Service
P.O. Box 41095
Craighall
Johannesburg 2024

SPAIN
Mundi-Prensa Libros, S.A.
Castello 37
28001 Madrid

Librería Internacional AEDOS
Consell de Cent, 391
08009 Barcelona

SRI LANKA AND THE MALDIVES
Lake House Bookshop
P.O. Box 244
100, Sir Chittampalam A. Gardiner Mawatha
Colombo 2

SWEDEN
For single titles
Fritzes Fackboksforetaget
Regeringsgatan 12, Box 16356
S-103 27 Stockholm

For subscription orders
Wennergren-Williams AB
Box 30004
S-104 25 Stockholm

SWITZERLAND
For single titles
Librairie Payot
6, rue Grenus
Case postal 381
CH 1211 Geneva 11

For subscription orders:
Librairie Payot
Service des Abonnements
Case postal 3312
CH 1002 Lausanne

TANZANIA
Oxford University Press
P.O. Box 5299
Dar es Salaam

THAILAND
Central Department Store
306 Silom Road
Bangkok

TRINIDAD & TOBAGO, ANTIGUA BARBUDA, BARBADOS, DOMINICA, GRENADA, GUYANA, JAMAICA, MONTSERRAT, ST. KITTS & NEVIS, ST. LUCIA, ST. VINCENT & GRENADINES
Systematics Studies Unit
#9 Watts Street
Curepe
Trinidad, West Indies

TURKEY
Haset Kitapevi, A.S.
Istiklal Caddesi No. 469
Beyoglu
Istanbul

UGANDA
Uganda Bookshop
P.O. Box 7145
Kampala

UNITED ARAB EMIRATES
MEMRB Gulf Co.
P.O. Box 6097
Sharjah

UNITED KINGDOM
Microinfo Ltd.
P.O. Box 3
Alton, Hampshire GU34 2PG
England

URUGUAY
Instituto Nacional del Libro
San Jose 1116
Montevideo

VENEZUELA
Libreria del Este
Aptdo. 60.337
Caracas 1060-A

YUGOSLAVIA
Jugoslovenska Knjiga
YU-11000 Belgrade Trg Republike

ZIMBABWE
Longman Zimbabwe
P.O. Box ST 125, Southerton
Harare

WHERE TO OBTAIN OECD PUBLICATIONS
OÙ OBTENIR LES PUBLICATIONS DE L'OCDE

Argentina - Argentine
Carlos Hirsch S.R.L.
Galeria Güemes, Florida 165, 4° Piso
1333 Buenos Aires
Tel. 30.7122, 331.1787 y 331.2391
Telegram: Hirsch-Baires
Telex: 21112 UAPE-AR. Ref. s/2901
Telefax:(1)331-1787

Australia - Australie
D.A. Book (Aust.) Pty. Ltd.
11-13 Station Street (P.O. Box 163)
Mitcham, Vic. 3132 Tel. (03)873.4411
Telex: AA37911 DA BOOK
Telefax: (03)873.5679

Austria - Autriche
OECD Publications and Information Centre
4 Simrockstrasse
5300 Bonn (Germany) Tel. (0228)21.60.45
Telex: 8 86300 Bonn
Telefax: (0228)26.11.04

Gerold & Co.
Graben 31
Wien I Tel. (0222)533.50.14

Belgium - Belgique
Jean De Lannoy
Avenue du Roi 202
B-1060 Bruxelles
Tel. (02)538.51.69/538.08.41
Telex: 63220 Telefax: (02)538.08.41

Canada
Renouf Publishing Company Ltd.
1294 Algoma Road
Ottawa, Ont. K1B 3W8 Tel. (613)741.4333
Telex: 053-4783 Telefax: (613)741.5439
Stores:
61 Sparks Street
Ottawa, Ont. K1P 5R1 Tel. (613)238.8985
211 Yonge Street
Toronto, Ont. M5B 1M4 Tel. (416)363.3171

Federal Publications
165 University Avenue
Toronto, ON M5H 3B9 Tel. (416)581.1552
Telefax: (416)581.1743
Les Publications Fédérales
1185 rue de l'Université
Montréal, PQ H3B 1R7 Tel.(514)954-1633

Les Éditions La Liberté Inc.
3020 Chemin Sainte-Foy
Sainte-Foy, P.Q. G1X 3V6
Tel. (418)658.3763
Telefax: (418)658.3763

Denmark - Danemark
Munksgaard Export and Subscription Service
35, Norre Sogade, P.O. Box 2148
DK-1016 Kobenhavn K
Tel. (45 33)12.85.70
Telex: 19431 MUNKS DK
Telefax: (45 33)12.93.87

Finland - Finlande
Akateeminen Kirjakauppa
Keskuskatu 1, P.O. Box 128
00100 Helsinki Tel. (358 0)12141
Telex: 125080 Telefax: (358 0)121.4441

France
OECD/OCDE
Mail Orders/Commandes par correspondance:
2 rue André-Pascal
75775 Paris Cedex 16 Tel. (1)45.24.82.00
Bookshop/Librairie:
33, rue Octave-Feuillet
75016 Paris Tel. (1)45.24.81.67
(1)45.24.81.81
Telex: 620 160 OCDE
Telefax: (33-1)45.24.85.00

Librairie de l'Université
12a, rue Nazareth
13602 Aix-en-Provence Tel. 42.26.18.08

Germany - Allemagne
OECD Publications and Information Centre
4 Simrockstrasse
5300 Bonn Tel. (0228)21.60.45
Telex: 8 86300 Bonn
Telefax: (0228)26.11.04

Greece - Grèce
Librairie Kauffmann
28 rue du Stade
105 64 Athens Tel. 322.21.60
Telex: 218187 LIKA Gr

Hong Kong
Government Information Services
Publications (Sales) Office
Information Service Department
No. 1 Battery Path
Central Tel. (5)23.31.91
Telex: 802.61190

Iceland - Islande
Mal Mog Menning
Laugavegi 18, Postholf 392
121 Reykjavik Tel. 15199/24240

India - Inde
Oxford Book and Stationery Co.
Scindia House
New Delhi 110001 Tel. 331.5896/5308
Telex: 31 61990 AM IN
Telefax: (11)332.5993
17 Park Street
Calcutta 700016 Tel. 240832

Indonesia - Indonésie
Pdii-Lipi
P.O. Box 269/JKSMG/88
Jakarta12790 Tel. 583467
Telex: 62 875

Ireland - Irlande
TDC Publishers - Library Suppliers
12 North Frederick Street
Dublin 1 Tel. 744835/749677
Telex: 33530 TDCP EI Telefax : 748416

Italy - Italie
Libreria Commissionaria Sansoni
Via Benedetto Fortini, 120/10
Casella Post. 552
50125 Firenze Tel. (055)645415
Telex: 570466 Telefax: (39.55)641257
Via Bartolini 29
20155 Milano Tel. 365083
La diffusione delle pubblicazioni OCSE viene assicurata dalle principali librerie ed anche da:
Editrice e Libreria Herder
Piazza Montecitorio 120
00186 Roma Tel. 679.4628
Telex: NATEL I 621427

Libreria Hoepli
Via Hoepli 5
20121 Milano Tel. 865446
Telex: 31.33.95 Telefax: (39.2)805.2886

Libreria Scientifica
Dott. Lucio de Biasio "Aeiou"
Via Meravigli 16
20123 Milano Tel. 807679
Telefax: 800175

Japan- Japon
OECD Publications and Information Centre
Landic Akasaka Building
2-3-4 Akasaka, Minato-ku
Tokyo 107 Tel. 586.2016
Telefax: (81.3)584.7929

Korea - Corée
Kyobo Book Centre Co. Ltd.
P.O. Box 1658, Kwang Hwa Moon
Seoul Tel. (REP)730.78.91
Telefax: 735.0030

Malaysia/Singapore - Malaisie/Singapour
University of Malaya Co-operative Bookshop Ltd.
P.O. Box 1127, Jalan Pantai Baru 59100
Kuala Lumpur
Malaysia Tel. 756.5000/756.5425
Telefax: 757.3661

Information Publications Pte. Ltd.
Pei-Fu Industrial Building
24 New Industrial Road No. 02-06
Singapore 1953 Tel. 283.1786/283.1798
Telefax: 284.8875

Netherlands - Pays-Bas
SDU Uitgeverij
Christoffel Plantijnstraat 2
Postbus 20014
2500 EA's-Gravenhage Tel. (070)78.99.11
Voor bestellingen: Tel. (070)78.98.80
Telex: 32486 stdru Telefax: (070)47.63.51

New Zealand -Nouvelle-Zélande
Government Printing Office
Customer Services
P.O. Box 12-411
Freepost 10-050
Thorndon, Wellington
Tel. 0800 733-406 Telefax: 04 499-1733

Norway - Norvège
Narvesen Info Center - NIC
Bertrand Narvesens vei 2
P.O. Box 6125 Etterstad
0602 Oslo 6
Tel. (02)67.83.10/(02)68.40.20
Telex: 79668 NIC N Telefax: (47 2)68.53.47

Pakistan
Mirza Book Agency
65 Shahrah Quaid-E-Azam
Lahore 3 Tel. 66839
Telex: 44886 UBL PK. Attn: MIRZA BK

Portugal
Livraria Portugal
Rua do Carmo 70-74
1117 Lisboa Codex Tel. 347.49.82/3/4/5

Singapore/Malaysia Singapour/Malaisie
See "Malaysia/Singapore"
Voir "Malaisie/Singapour"

Spain - Espagne
Mundi-Prensa Libros S.A.
Castello 37, Apartado 1223
Madrid 28001 Tel. (91) 431.33.99
Telex: 49370 MPLI Telefax: (91) 275.39.98

Libreria Internacional AEDOS
Consejo de Ciento 391
08009 -Barcelona Tel. (93) 301-86-15
Telefax: (93) 317-01-41

Sweden - Suède
Fritzes Fackboksföretaget
Box 16356, S 103 27 STH
Regeringsgatan 12
DS Stockholm Tel. (08)23.89.00
Telex: 12387 Telefax: (08)20.50.21
Subscription Agency/Abonnements:
Wennergren-Williams AB
Box 30004
104 25 Stockholm Tel. (08)54.12.00
Telex: 19937 Telefax: (08)50.82.86

Switzerland - Suisse
OECD Publications and Information Centre
4 Simrockstrasse
5300 Bonn (Germany) Tel. (0228)21.60.45
Telex: 8 86300 Bonn
Telefax: (0228)26.11.04

Librairie Payot
6 rue Grenus
1211 Genève 11 Tel. (022)731.89.50
Telex: 28356

Maditec S.A.
Ch. des Palettes 4
1020 Renens/Lausanne Tel. (021)635.08.65
Telefax: (021)635.07.80

United Nations Bookshop/Librairie des Nations-Unies
Palais des Nations
1211 Genève 10
Tel. (022)734.60.11 (ext. 48.72)
Telex: 289696 (Attn: Sales)
Telefax: (022)733.98.79

Taïwan - Formose
Good Faith Worldwide Int'l. Co. Ltd.
9th Floor, No. 118, Sec. 2
Chung Hsiao E. Road
Taipei Tel. 391.7396/391.7397
Telefax: (02) 394.9176

Thailand - Thalande
Suksit Siam Co. Ltd.
1715 Rama IV Road, Samyan
Bangkok 5 Tel. 251.1630

Turkey - Turquie
Kültur Yayinlari Is-Türk Ltd. Sti.
Atatürk Bulvari No. 191/Kat. 21
Kavaklidere/Ankara Tel. 25.07.60
Dolmabahce Cad. No. 29
Besiktas/Istanbul Tel. 160.71.88
Telex: 43482B

United Kingdom - Royaume-Uni
H.M. Stationery Office
Gen. enquiries Tel. (01) 873 0011
Postal orders only:
P.O. Box 276, London SW8 5DT
Personal Callers HMSO Bookshop
49 High Holborn, London WC1V 6HB
Telex: 297138 Telefax: 873.8463
Branches at: Belfast, Birmingham, Bristol, Edinburgh, Manchester

United States - États-Unis
OECD Publications and Information Centre
2001 L Street N.W., Suite 700
Washington, D.C. 20036-4095
Tel. (202)785.6323
Telex: 440245 WASHINGTON D.C.
Telefax: (202)785.0350

Venezuela
Libreria del Este
Avda F. Miranda 52, Aptdo. 60337
Edificio Galipan
Caracas 106
Tel. 951.1705/951.2307/951.1297
Telegram: Libreste Caracas

Yugoslavia - Yougoslavie
Jugoslovenska Knjiga
Knez Mihajlova 2, P.O. Box 36
Beograd Tel. 621.992
Telex: 12466 jk bgd

Orders and inquiries from countries where Distributors have not yet been appointed should be sent to: OECD Publications Service, 2 rue André-Pascal, 75775 Paris Cedex 16.

Les commandes provenant de pays où l'OCDE n'a pas encore désigné de distributeur devraient être adressées à : OCDE, Service des Publications, 2, rue André-Pascal, 75775 Paris Cedex 16.

1/90

Printed on behalf of the World Bank
and the Organisation for Economic Co-operation and Development by
OECD PUBLICATIONS 2, rue André-Pascal 75775 PARIS CEDEX 16
PRINTED IN FRANCE
(77 89 05 1) ISBN 92-64-13309-7 No. 44923 1990
ISBN 0-8213-1391-6